D0812594

PPh

Progress in Physics

Edited by A. Jaffe and D. Ruelle

ITERATED MAPS ON THE INTERVAL AS DYNAMICAL SYSTEMS

Pierre Collet

Jean-Pierre Eckmann

BIRKHÄUSER

BASEL · BOSTON · STUTTGART

Authors

Pierre Collet

Ecole Polytechnique
Palaiseau
France

Jean-Pierre Eckmann

Département de Physique Théorique
Université de Genève
CP-1211 Genève 4
Switzerland

Library of Congress Cataloging in Publication Data

Collet, Pierre, 1948—
Iterated maps on the interval as dynamical systems.

(Progress in physics ; 1)
Bibliography: p.
Includes index.
1. Differentiable dynamical systems. 2. Mappings (Mathematics) I. Eckmann, Jean Pierre, joint author. II. Title. III. Series: Progress in physics (Boston) ; 1.
QA614.8.C64 003 80-20751
ISBN 3-7643-3026-0

CIP—Kurztitelaufnahme der Deutschen Bibliothek

Collet, Pierre:
Iterated maps on the interval as dynamical systems / Pierre Collet : Jean-Pierre Eckmann.
- Boston, Basel, Stuttgart : Birkhäuser, 1980.
(Progress in physics : 1)
ISBN 3-7643-3026-0

NE: Eckmann, Jean-Pierre:

ISBN: 3-7643-3026-0

Printed in USA

TABLE OF CONTENTS

Relations between the sections

Logical dependence : ⟶ ([→] only partially)

Related subjects : Part I ←----→ Parts II, III

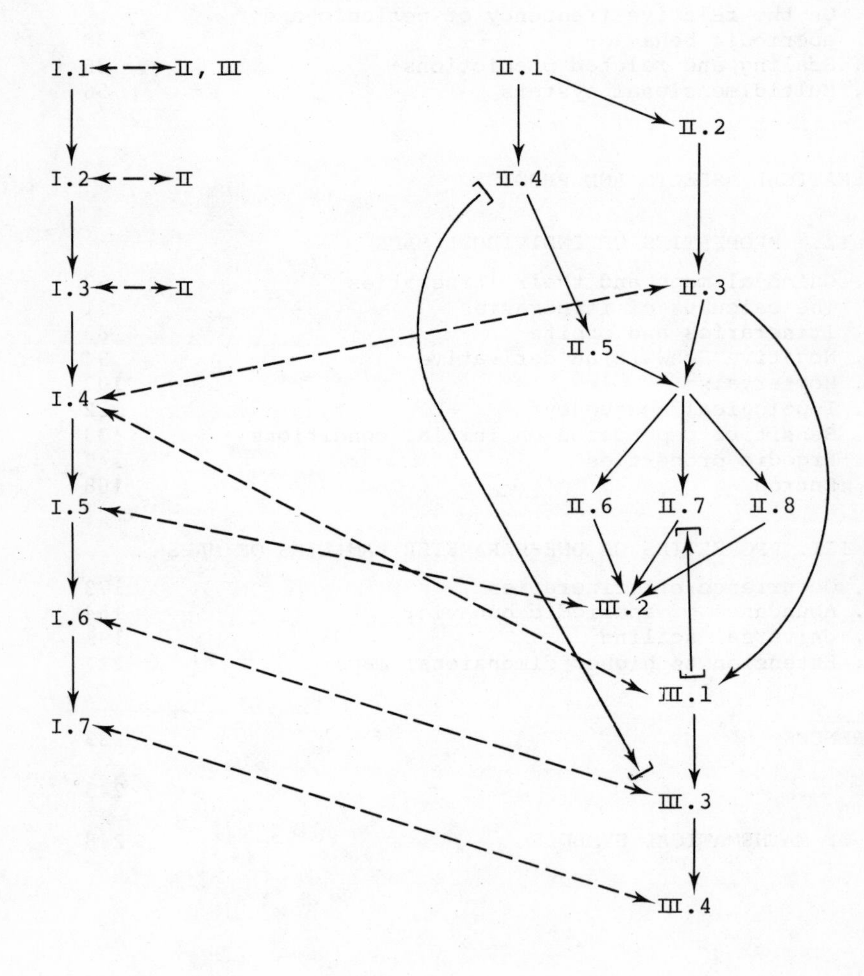

INTRODUCTION

Iterations of continuous maps of an interval into itself serve as the simplest examples of models for dynamical systems. These models present an interesting mathematical structure going far beyond the simple equilibrium solutions one might expect. If in addition the dynamical system depends on an experimentally controllable parameter, there is a corresponding mathematical structure telling a lot about interrelations between the behavior for different parameter values.

The purpose of this work is to explain some of the results of this theory to mathematicians and theoretical physicists, with the additional hope of stimulating experimentalists (on computers or in Nature) to look for more of these general phenomena of beautiful regularity, which seem always to appear near to the much less understood chaotic regimes. The word "Physics" is used throughout to denote the natural sciences in general. Continuous maps of an interval to itself seem to have been introduced first to model biological systems. Now they can be found as models in most natural sciences and we have even heard of such models for economics.

One cannot write about the subject of maps without reference to the beautiful review article by May which appeared in 1976. This review has been very stimulating for further research. The main reason for writing another review about maps is the progress made in this field in the meantime, and the associated shift of emphasis to new questions. Another reason is that no review has, to our knowledge, been written about the mathematics of the subject.

The organization of this book is as follows. There are three parts: Part I. Motivation and Interpretation; Parts II, III. Mathematical Results. They are written in such a way that the reader can read Part I alone, or Parts II, III alone, or both in either order. If a reader finds Part I too mathematical, a prior reading of May's article could be helpful.

The proofs in Parts II, III are adaptations of those found in the literature, usually simplified by restricting somewhat the class of functions to which they apply.

It should be emphasized already in this introduction that we consider the study of one-dimensional maps as much more than just a mathematical curiosity. It should rather be seen as an attempt to isolate simplifying features of dissipative dynamical systems, i.e., of physical systems with some sort of "friction." Such systems tend in general to equilibrium positions when left alone. But when they are driven from the outside, they quite often show behavior which is called erratic, aperiodic, turbulent, or strange. This behavior is very ill understood. For example, one is far from a complete classification of different types of motion and the interrelationships between them. One reason for studying continuous maps of the interval is thus their simplicity. We shall see that the orbit structure is fairly well understood for a relatively large class of maps (Sections II.1-4). Next we study some typical possible types of behavior which occur for these maps, of increasingly aperiodic nature (Sections II.5-8). In Part III we study one parameter families of maps. The parameter should be thought of as some controllable quantity. The main results we describe are: The dependence of the orbit structure on the parameter, the abundance of aperiodic behavior in certain situations. Most important for applications, we exhibit a kind of universal behavior for the successive bifurcations in one parameter families of maps. This seems so important to us, that we describe in a special section the extension of the result to maps of $\mathbb{R}^n$. The relevance of this extension is illustrated in Part I, Sections 1, 3, 6, 8, where we illustrate the connections with hydrodynamical experiments. We believe that the scaling predictions which can be made in relation to the study of maps on the interval (and of $\mathbb{R}^n$) will become directly relevant in many concrete systems. But also the relations between periodic and aperiodic behavior which are relatively clear for the case of maps on the interval should help us to understand at least some parts of the more complicated multidimensional case. Thus we hope that this book helps to open some new perspectives on dynamical systems.

Substantial parts of this book describe work of J. Guckenheimer and M. Misiurewicz, and other sections deal with work we have done in collaboration with H. Koch and O. Lanford. We hope that our effort to give a coherent overview of this work will justify our extensive use of these sources. We have followed them often rather closely, presenting some new proofs where it seemed adequate.

We are very grateful to L. Jonker and D. Rand for their detailed constructive criticism of our manuscript which led to several improvements. We are also indebted to D. Ruelle, who, over the years, has patiently explained the concepts of dynamical systems and their relevance to physics. This, and the stimulating atmosphere at Bures-sur-Yvette are mainly responsible for the genesis of this book, but the initial stimulus is due to A. Jaffe.

We thank R. D'Arcangelo and F. Nicole for their excellent and rapid typing.

Various institutions have made our collaboration possible : IHES (Bures-sur-Yvette), NSF (Grant Phy 79-16812 and French-American Seminar on Statistical mechanics), Fonds National Suisse. In addition, we have profited from the computing equipment of the DPMC at Geneva University in doing our research and in preparing the drawings of this book. We are grateful for this support.

PART I. MOTIVATION AND INTERPRETATION

I.1 ONE-PARAMETER FAMILIES OF MAPS

1. The iterations of maps of an interval into itself certainly present one of the easiest models or examples of nonlinear (dissipative) dynamical systems. In fact, iterations of the form $x_n \to x_{n+1} = f(x_n)$, where f maps $[-1,1]$ into itself, can be viewed as a discrete time version of a continuous dynamical system as we shall see below. Here, n plays the role of the time variable. Such iterations have been advocated with more or less success as models for biological, chemical or physical systems. We shall take here a more general point of view by asking questions which are not related to any specific map, but whose answer is, to some extent, independent of the particular map. While we lose, of course, any detailed information, we shall see that a lot of qualitative, and even more surprisingly, some quantitative results hold in a very wide context. We would like to stress that this does not mean that modelling in science can be done carelessly but it tells us that the mere fact of adopting a certain type of modelling, e.g., choosing to describe the situation in discrete time, has already a profound influence on the outcome of the analysis. So a new type of question, different from what is taught in general in science curricula, may become relevant. One of the aims of this book is to show which aspects of dynamical systems give rise to such questions.

2. It might seem that the choice of one-dimensional dynamical systems is unduly restrictive. This is true in the sense that multidimensional systems may show phenomena which simply cannot occur in one dimension due to the restricted volume of "phase space" (the interval $[-1,1]$). Furthermore, the natural ordering of the real numbers has profound combinatorial influences on the possible behavior of iterated maps on the interval. On the other hand, as we shall see, some results extend straightforwardly to $\mathbb{R}^n$.

Many physical systems are modelled on a continuous time through a differential equation of the form

$$\frac{d}{dt} x(t) = F(x(t)) \tag{1}$$

where $x \in \mathbb{R}^n$ and $F: \mathbb{R}^n \to \mathbb{R}^n$ is a function on the phase space $\mathbb{R}^n$ (the force field). There are then two ways to relate such an equation to an iteration of a map of $\mathbb{R}^m$ into itself. The easier, but less rigorous way consists in discretizing time. This leads to the equation (for any choice of (small) $t_0 > 0$),

$$x((n+1)t_0) \sim x(nt_0) + t_0 F(x(nt_0)) ,$$

or

$$x_{n+1} = x_n + t_0 F(x_n) = f(x_n) , \text{ with } x_n = x(nt_0),$$

which is of the desired form.

The other method can be applied if Eq. (1) has a periodic solution, i.e., if for some initial condition $x(0) = x_0$ we find $x(T) = x(0)$ for some $T > 0$. Then we consider a hyperplane $\mathbb{R}^{n-1}$ of dimension $n-1$ transverse to the curve $t \to x(t)$ through x_0, and in this hyperplane a (small) neighborhood $\mathcal{U}$ of x_0.

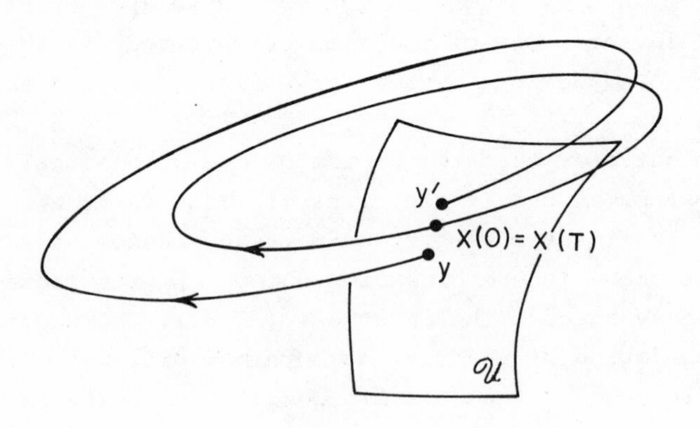

Figure I.1.

Then a map $P: \mathcal{U} \to \mathbb{R}^{n-1}$ is induced by associating to $y \in \mathcal{U}$ the next intersection with $\mathbb{R}^{n-1}$ of the trajectory with initial condition $x(0) = y$. If the first such intersection occurs at y' we define $Py = y'$. (Note that the time needed to go from y to y' may be different from T, and some points might never return, a case we do not consider.) P is then a map $\mathcal{U} \to \mathbb{R}^{n-1}$, and this is how flows reduce sometimes to maps. Given a map $P: \mathbb{R}^{n-1} \to \mathbb{R}^{n-1}$, it may still be possible to reduce the discussion further (at least in part) if the map P leaves invariant a set of "parallel" hypersurfaces of dimension $n-2$ in $\mathbb{R}^{n-1}$. That is, it maps one hyperplane

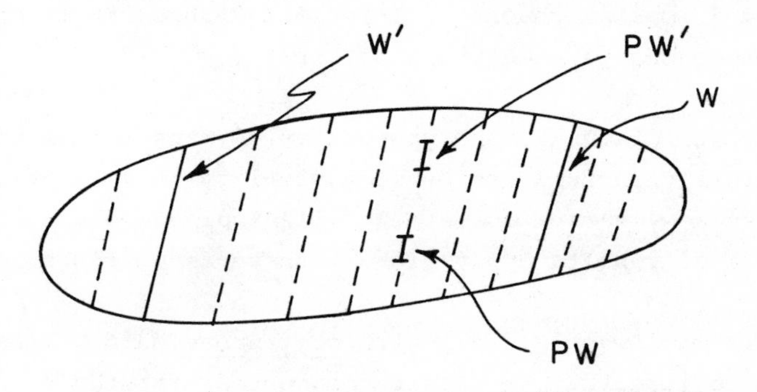

Figure I.2.

onto another hyperplane. Using each plane as a coordinate, we are reduced to a one-dimensional problem. Information about this one-dimensional problem or about P will give us some insight about the behavior of the flow itself.

3. In many applications of natural sciences, the dynamical system depends on controllable quantities, which we call <u>parameters</u> for short. Examples are temperature, Reynolds number or energy fed into a system. A system may depend on many parameters, but we shall see that in fact every one-parameter family of one-dimensional maps of a suitable type will go through the whole spectrum of possibilities. Therefore, in one dimension it is sufficient to study one-parameter families of maps, since no new phenomena can be found by using more parameters.

4. If we are given a one-parameter family of maps, parametrized by $\mu \in \mathbb{R}$ there is still another variable quantity, namely the choice of initial point $x_0 \in [-1,1]$. We shall be interested in the behavior of a "typical" initial point for a "typical" value of the parameter. Of course, appropriate notions will be developed below. But even in this informal introduction we want to stress the importance of looking for typical behavior. As we shall see, even for very well behaved maps there may be some "atypical" initial points for which something chaotic happens, but such initial points are extremely rare in the measure theoretic sense. In particular the title "Period 3 implies chaos " Li-Yorke [1975] has, in the past, led to some confusion about this matter.

5. The discussion of properties of maps on the interval is extremely streamlined by some simplifying assumptions of which we now mention a few. We will indicate whenever extensions of results apply without these assumptions.

This review only deals with continuous maps of the interval $[-1,1]$ to itself. The results connected with the natural ordering of the real numbers break down when the continuity is abandoned. The choice of the interval $[-1,1]$ is of course arbitrary, since the change of coordinates $x = (2x'-(b+a))/(b-a)$ will transform a map $[a,b] \to [a,b]$ into a map $[-1,1] \to [-1,1]$. If $a = -\infty$ and/or $b = +\infty$ we are in a truly different situation, which can however often be reduced to the situation $[-1,1]$ by the following type of discussion. Assume, e.g., $a = 0$, $b = \infty$ and $f(x) \to c$ with $|c| < \infty$ as $x \to +\infty$. Then the set $[0,\infty]$ is mapped onto $[\inf_{x\geq 0} f(x), \sup_{x\geq 0} f(x)] = [a',b']$ and we are reduced to the case of finite $[a,b]$ if we consider instead of initial points $x \in [0,\infty]$ their images $f(x)$ as initial points.

On the other hand, if $f(x)$ tends to infinity as $x \to \infty$, there may be a set of points U, such that $f^n(x) \to \infty$ as $n \to \infty$ for all x in U. Thus U is the set of points which escape to infinity. It may happen that the complement of U is an interval, and in this case we may be reduced to the previous discussion.

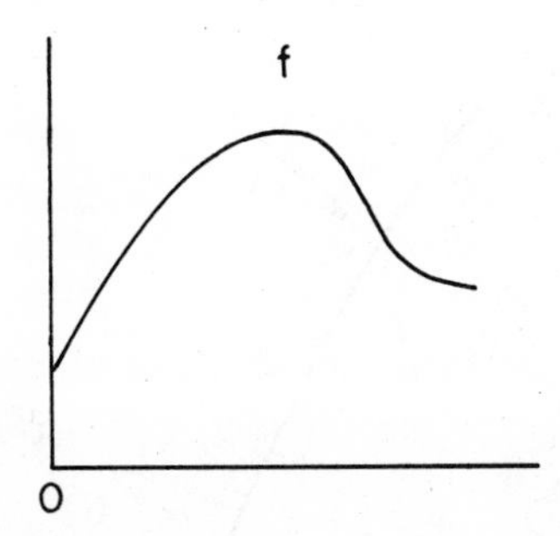

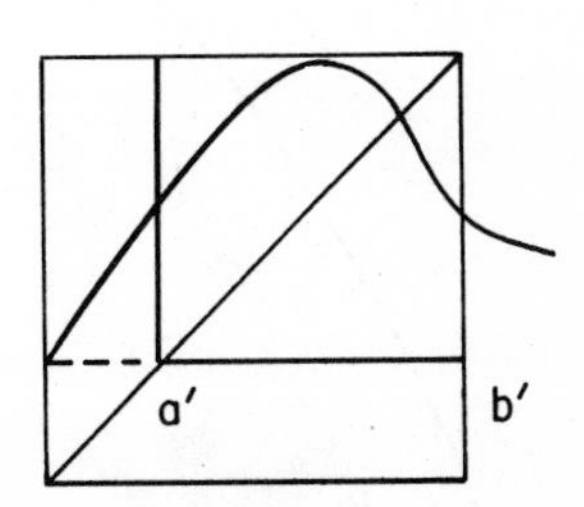

Figure I.3.

We shall consider only maps with <u>one maximum</u> and we assume that this maximum occurs for $x = 0$. This can sometimes be achieved by a differentiable coordinate transformation. We further assume f is monotonically increasing for $x < 0$, and monotonically decreasing for $x > 0$, and once continuously differentiable (see II.1, for details). A very useful simplifying assumption which we shall make throughout is the convexity statement of "negative Schwarzian derivative," namely that $|f'|^{-1/2}$ is a convex function on $x < 0$ and on $x > 0$ (cf. II.4).

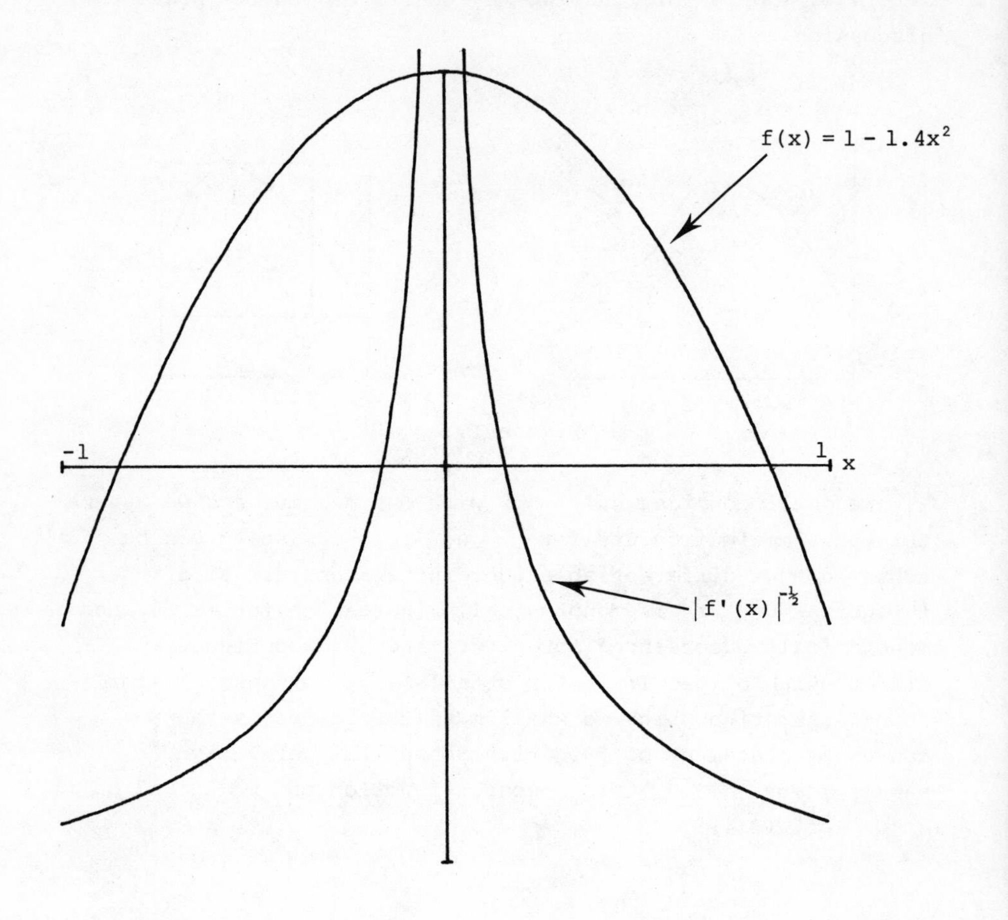

Figure I.4. A typical function f, and the graph of $|f'|^{-1/2}$.

I.2 TYPICAL BEHAVIOR FOR ONE MAP

Before we study parametrized families of maps, we want to analyze individual maps. We are interested in the possible behavior of the successive images of an initial point x_0 on the interval $[-1,1]$ for a fixed map f. For this we first outline a graphical method for determining the iterates $x_n = f^n(x_0)$. Here, we define $f^n(x_0) = f(f^{n-1}(x_0))$. The following Figure I.5 shows how this is done through the rule: Go from x_0 to the graph of the function, from the graph to the diagaonal, from the diagonal to the graph,... .

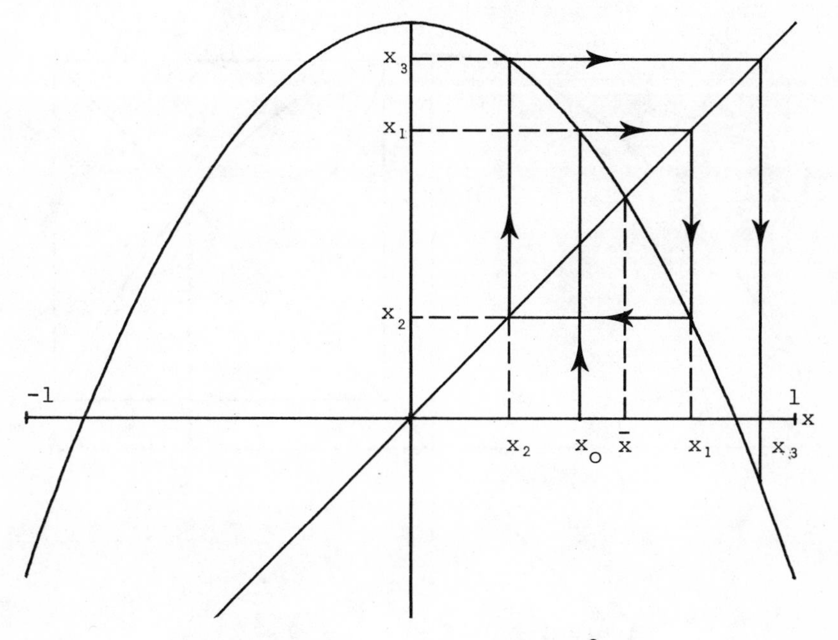

Figure I.5. $f(x) = 1 - 1.4\,x^2$

Note that the point marked $\bar{x}$ does not move at all; $f^n(\bar{x}) = f(\bar{x}) = \bar{x}$. We shall call $\bar{x}$ a <u>fixed point</u> of f. Physically speaking, this means that if the system is at $\bar{x}$ at some time, it will remain there forever. Going back to our continuous systems the Poincaré map will have a fixed point, if the system has a closed (and hence periodic) orbit. Such orbits are sometimes called limit cycles.

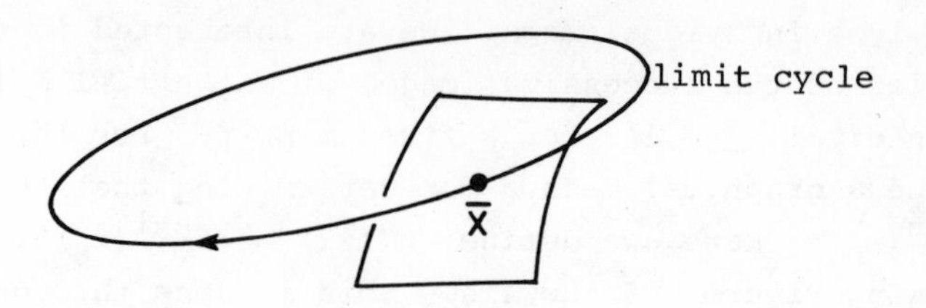

Figure I.6.

Let us now consider another map.

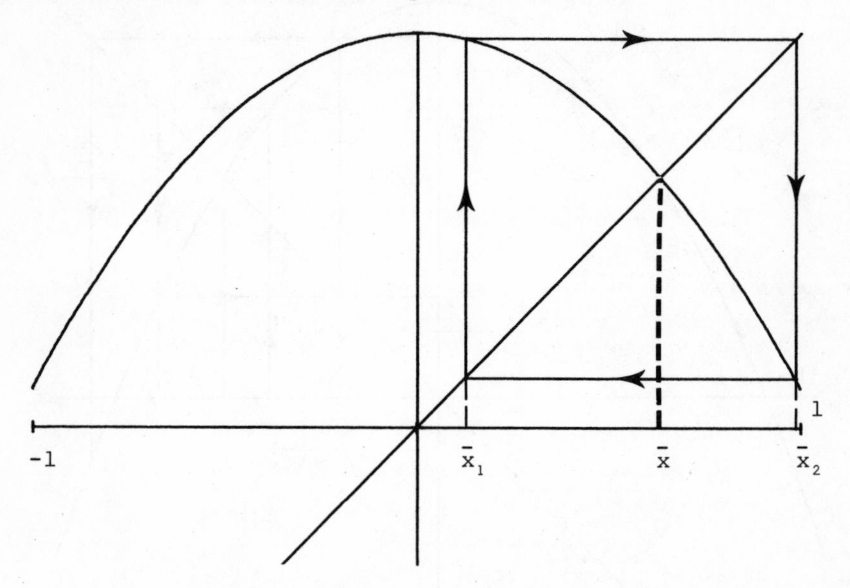

Figure I.7. $f(x) = 1 - .9\,x^2$.

This map has the property that the point $\bar{x}_1$ satisfies $f(\bar{x}_1) = \bar{x}_2$ and $f(\bar{x}_2) = \bar{x}_1$ or in other terms $f^2(\bar{x}_1) = \bar{x}_1$, and $f^2(\bar{x}_2) = \bar{x}_2$. One says f has a periodic orbit of period 2 (which is implied by the fact that f^2 has two fixed points namely $\bar{x}_1, \bar{x}_2$ which are not fixed points for f).

Again, we have an analogous picture for flows.

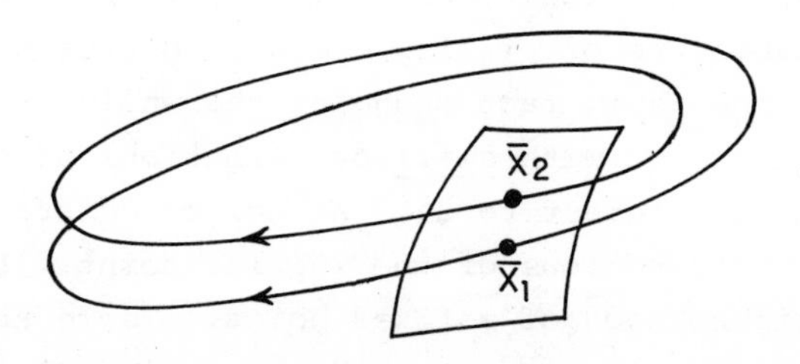

Figure I.8.

We come now to a very important point of the discussion. If a fixed point (or a periodic point) is to be relevant for observations in the dynamical system described by f, we must see whether it persists under small perturbations. There are different kinds of perturbations envisageable, namely

(1) perturbation of the intial point: "Do systems in nearby initial states evolve similarly?"

(2) perturbations of the function f: "f is only approximately known."

(3) stochastic perturbations: "The true equation is not $x_n = f(x_{n-1})$ but there will be noise terms which can be modelled by saying that $x_n = f(x_{n-1}) + r(x_{n-1})$ where $r(x)$ is a small random step, i.e., a variation of $f(x_{n-1})$ with some a priori probability distribution."

We shall analyze below in great detail the Case (1). The motivations for this are two-fold: (A) There are some situations where, if (1) is under control, then (2) and (3) do not have an interesting effect: i.e., small perturbations of f or small random steps do not affect the qualitative behavior of the system. (B) There are situations where (2) and (3) are ill-understood and no results are available, (cf. Kifer [1974] for a positive result in this direction).

However, we insist that considering large random forces or large perturbations of f is a totally different enterprise, because one is in fact changing the whole problem, and this has nothing to do with small perturbations of the system, which are of main interest to us. We now concentrate, as announced, on perturbations of the initial point. Let us thus analyze the neighborhood of a fixed point. There are two basic situations, with regards to the long-time (i.e., f^n for n large) evolution of a point near a fixed point.

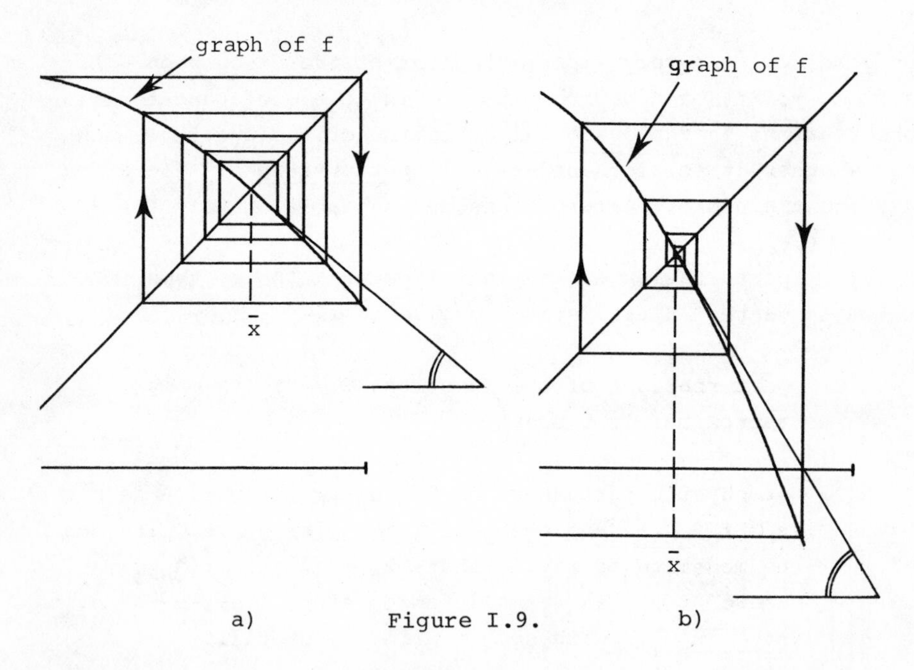

Figure I.9.

We see that if the slope $f'(x)$ of f at $\bar{x}$ satisfies $|f'(\bar{x})| < 1$ (angle $<45°$ with horizontal, Fig. I.9A) then a point x_0 near $\bar{x}$ will have the property that $\lim_{n\to\infty} f^n(x_0) = \bar{x}$ and in fact this is true for all choices of x_0 in a sufficiently small neighborhood of $\bar{x}$. If, on the other hand, as in Fig. I.9B, $|f'(\bar{x})| > 1$, then there is a neighborhood $\mathcal{U}$

of $\bar{x}$ such that $f^{n_0}(x_0) \notin \mathcal{U}$ after some number n_0 of steps, when $x_0 \neq \bar{x}$. The number n_0 of steps depends on x_0 and on the neighborhood. Note that the global form of f can be such that $f^{n_0+k}(x_0)$ is again in the neighborhood $\mathcal{U}$ for some later k, depending on x_0 but this is not of concern to us now. In the case $|f'(\bar{x})| < 1$ we call $\bar{x}$ a <u>stable fixed point</u>, in the opposite case $\bar{x}$ is called an <u>unstable fixed point</u>. The case of $|f'(\bar{x})| = 1$ will be dealt with in Section II.4. The same kind of analysis can be made for periodic points. E.g., in the case of a periodic point $\bar{x}_1$ of period 2, we saw that $f^2(\bar{x}_1) = \bar{x}_1$, so we get a fixed point if we consider f^2 instead of f. Thus the condition for stability becomes

$$|f^{2'}(\bar{x}_1)| \leq 1 \quad .$$

By the chain rule of differentiation,

$$f^{2'}(\bar{x}_1) = f'(f(\bar{x}_1))f'(\bar{x}_1) = f'(\bar{x}_1)f'(\bar{x}_2)(=f^{2'}(\bar{x}_2)),$$

i.e., the derivative is the <u>product</u> of derivatives along the periodic orbit. We give an example of a stable periodic orbit of period 2.

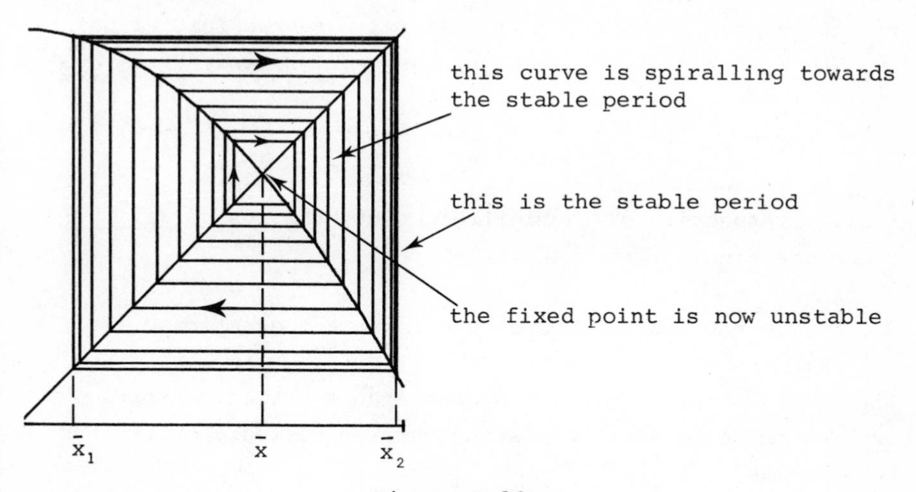
this curve is spiralling towards the stable period
this is the stable period
the fixed point is now unstable
$\bar{x}_1$
$\bar{x}$
$\bar{x}_2$

Figure I.10.

Of course, all this may be easily translated to the case of periodic orbits of arbitrary finite length. Also, we can pass again to flows, and speak of stable limit cycles and unstable limit cycles.

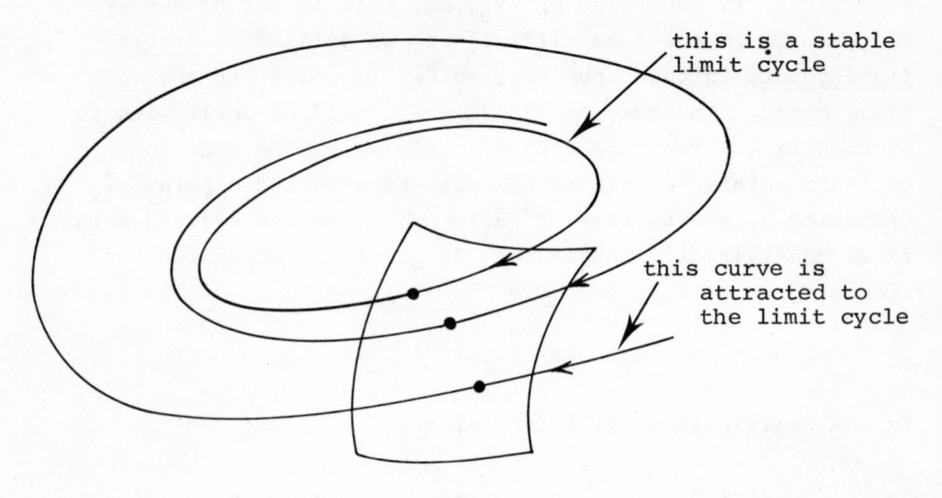
this is a stable limit cycle
this curve is attracted to the limit cycle

Figure I.11.

In accordance with the point of view (1), we see that stable periodic orbits are relevant for physical systems, because many initial points will eventually show the same behavior for large n. Namely, if x_0 and x'_0 are two initial points near to a stable fixed point $\bar{x}$, they will satisfy

$$\lim_{n\to\infty} f^n(x_0) = \lim_{n\to\infty} f^n(x'_0) = \bar{x} ,$$

i.e., irrespective of the initial point, the system will reach the final state $\bar{x}$.

We can now ask and answer two important questions.

Q1: Do all points converge to some stable periodic orbit?
Q2: Can there be several distinct stable periodic orbits for one map?

The answer to Q2 is, under the hypotheses of negative Schwarzian derivative:

TH1 There can be at most one stable periodic orbit (II.4.2).

Note the "at most"! We shall see later that there are maps (in fact many) which have no stable periodic orbit. We shall call them aperiodic maps. (But all continuous maps from $[-1,1]$ to itself have at least one unstable or stable fixed point). For the moment, we shall concentrate on those maps with a stable periodic orbit. Then it is reasonable to ask question Q1. The answer is, of course, "no". Not all points converge to a stable periodic orbit. It suffices to look at Fig. I.10. There we have a stable periodic orbit $\bar{x}_1, \bar{x}_2$ but we also have the point $\bar{x}$, defined by $f(\bar{x}) = \bar{x}$: obviously it is an unstable fixed point (slope >45°, and TH1 above) but the point $x_0 = \bar{x}$ will satisfy $f^n(x_0) = \bar{x}$ and hence it does not tend to $\bar{x}_1, \bar{x}_2$. So we see that the "no" to question Q1 is unavoidable for many maps. But there is a more reasonable alternative question.

Q1': How many points converge to the stable periodic orbit?

with the answer:

TH2 The measure of those points which do not converge to the stable periodic orbit is zero. (II.5.7)

Measure is Lebesgue measure, and if you are not familiar with this concept, you can replace the statement by the weaker one that those points which do not converge to the stable periodic orbit do not form subintervals of the line $[-1,1]$.

How can we determine whether a given map f has a stable period or not? First of all, this question should not suggest that by looking at the analytic form of f (even if $f(x) = 1-\mu x^2$) one can decide on the occurrence or absence of a stable periodic orbit. This is in fact a hopeless problem.

We should rather ask how to find all the stable periodic orbits whose periods are not too long. One has the following criterion.

TH3 <u>If</u> f <u>has a stable periodic orbit, then the initial point</u> 0 <u>will be attracted to it</u>. (II.4.2)

(Recall that 0 is the point for which $f' = 0$.) In other words, 0 never belongs to the exceptional set described in the preceding discussion. It is now legitimate to ask whether this criterion is ever useful. In fact it is--at least for short periods--and furthermore, it provides us with a tool to construct a <u>map without a stable periodic orbit</u>. This map is the map $x \to 1 - 2x^2$ (Fig. I.12).

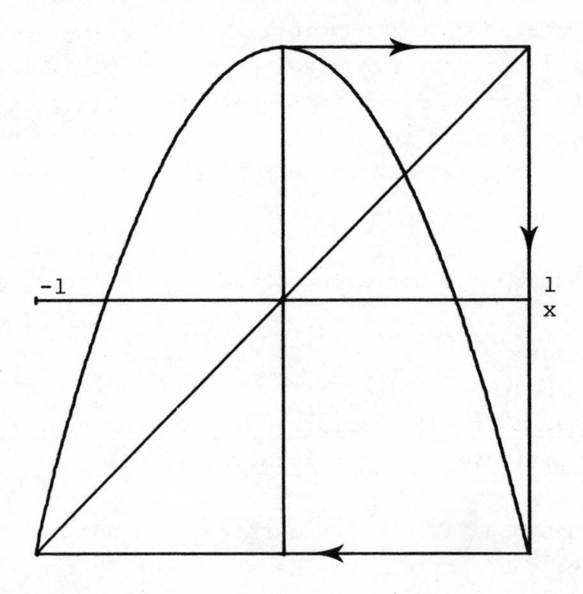

Figure I.12. $f(x) = 1 - 2x^2$

If we take $x_0 = 0$, then $x_1 = f(x_0) = 1$ and $x_2 = f^2(x_0) = -1$ and then $x_n = -1$ for all $n \geq 2$. Therefore the image of $x_0 = 0$ "settles down" after 2 iterations. But -1 (which is obviously a fixed point of the map f) is <u>not a stable fixed point</u>. Assume that f has a stable periodic orbit somewhere.

Then the images of 0 will not be attracted to it, because they stay at -1, and -1 is not part of the periodic orbit, since it is a fixed point. We arrive thus at a contradiction and it follows that f does not have a stable periodic orbit.

It is now natural to ask a new question. What happens to a typical initial point when there is no stable periodic orbit? Two essential cases have been studied, but it is not known whether there are other typical cases.

The first case occurs for $f(x) = 1 - 2x^2$, the case we have just examined. Let us perform the following experiment. We take as initial point x_0 "any" point (not $x_0 = 0$), and iterate it 50000 times. Then we plot the histogram for the number of points which have fallen in each of the 200 intervals $[n/100,(n+1)/100])$, $n = -100,-99,\ldots,99$. The result is shown in Fig. I.13. This curve is very near to $(\pi(1-y^2)^{1/2})^{-1}$, see Figure I.14. [The map $f(x) = 1 - 2x^2$ was studied by Ulam and von Neumann [1947]. If one takes as new coordinates

$$y = \frac{4}{\pi}\left(\arcsin\sqrt{\frac{x+1}{2}}\right) - 1 \quad ,$$

then, in these new coordinates, f takes the form $\hat{f}(y) = 1 - 2|y|$. This function has obviously no stable periodic orbits (the derivative is everywhere ± 2).]

The kind of behavior exhibited by the function f is called ergodic behavior. It says, in essence, that most points visit every region of phase space with about equal probability. But in fact, our example is much more chaotic in the following sense. Consider two nearby points x_0 and x_0' and their respective evolutions $x_n = f^n(x_0)$, $x_n' = f^n(x_0')$. For our function f, in general, no matter how near x_0 is to x_0', for some n, the points x_n and x_n' will eventually be noticeably separated. Ruelle [1978(1)] has coined the term sensitive dependence on initial conditions for this kind of behavior (II.7). It explains in a very appealing way the apparent incompatibility

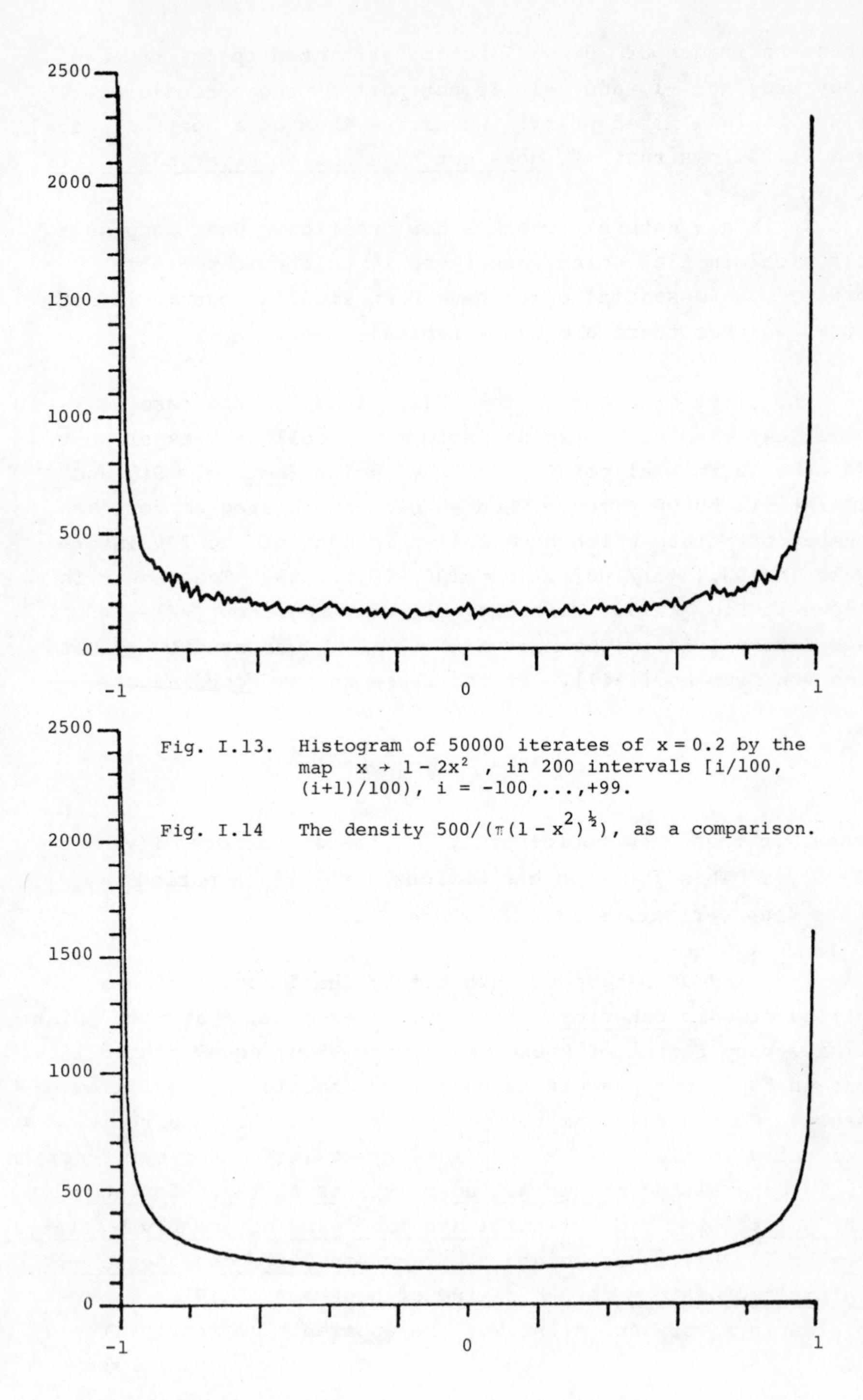

Fig. I.13. Histogram of 50000 iterates of $x = 0.2$ by the map $x \to 1 - 2x^2$, in 200 intervals $[i/100, (i+1)/100)$, $i = -100,\dots,+99$.

Fig. I.14 The density $500/(\pi(1-x^2)^{\frac{1}{2}})$, as a comparison.

between the determinacy of a system, and the unpredictability of its time evolution. In fact, any imprecision of our knowledge of x_0, no matter how small, will eventually show up on the scale of the interval. Furthermore, this amplification of error can be quite violent and rapid, and for our previously discussed example, each iteration amplifies the error by two, since the derivative of $1-2|x|$ equals ± 2, and

$$\hat{f}(x_0) - \hat{f}(x_0') \sim \hat{f}'(x_0)(x_0-x_0') = \pm 2(x_0-x_0') \quad .$$

Another nice way to say this has been illustrated by Shaw [1978]: One can view the sensitivity to initial conditions as forgetting where a point comes from. Let us perform the following game to see this. For the map $x \to 1-2x^2$ we choose 10000 initial points $x_0^{(1)},\ldots,x_0^{(10000)}$ near $-1/4$, (more precisely $x_0^{(m)} = -1/4 + (m-1)10^{-15}$) and we ask how many of the points $x_n^{(m)} = f^n(x_0^{(m)})$ have not hit the "target" where we define the target to be the interval $(-0.22, -.2)$. If they hit it, they are "dead" and we pursue only the fate of the "survivors ".

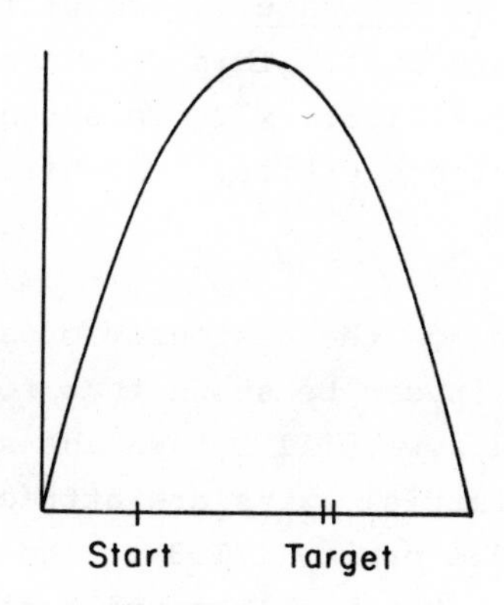

Figure 1.15

We obtain an exponential

$$\text{survivors} \sim 10000 \exp(-n/\tau) ,$$

where the lifetime τ is simply given by the probability of hitting the target when the points are distributed according to the histogram, Fig. I.13. Since the theoretical distribution is $P(y) = 1/(\pi\sqrt{1-y^2})$ we find

$$\tau \sim (0.02P(x))^{-1} = 50/\pi \sqrt{1-0.2^2} \sim 154$$

iterations. This fits the observed slope very well. On the other hand, since the precision of our information decreases by a factor of 2 per iteration, we shall have totally forgotten the initial data after about

$$-\log(10^{-15} \cdot 10^4)/\log 2 \sim 36$$

iterations provided the initial interval does not hit the target right away. So during the first 36 iterations, the information is well retained and then an exponential falloff can be observed. See Figure I.16.

We come now to a second case (which will turn out to be quite rare but very crucial). This case occurs for the function $f(x) = 1 - 1.401155\ldots x^2$. This function can be shown to have no stable periodic orbits. The histogram looks like in Figure I.17.

Note that this is not the histogram of a long stable periodic orbit. But it can be shown that for almost all initial choices of x_0 we will obtain the same histogram. That is, almost all initial x_0's are attracted to the same stable, but nonperiodic orbit (III.3). The volume occupied by this orbit is of (Lebesgue-) measure zero, i.e., the orbit occupies no length (volume). This is in contrast to the first case (of the function $f(x) = 1 - 2x^2$). Furthermore, there is another marked difference: The map in question does not have sensitive dependence with respect to initial conditions. In

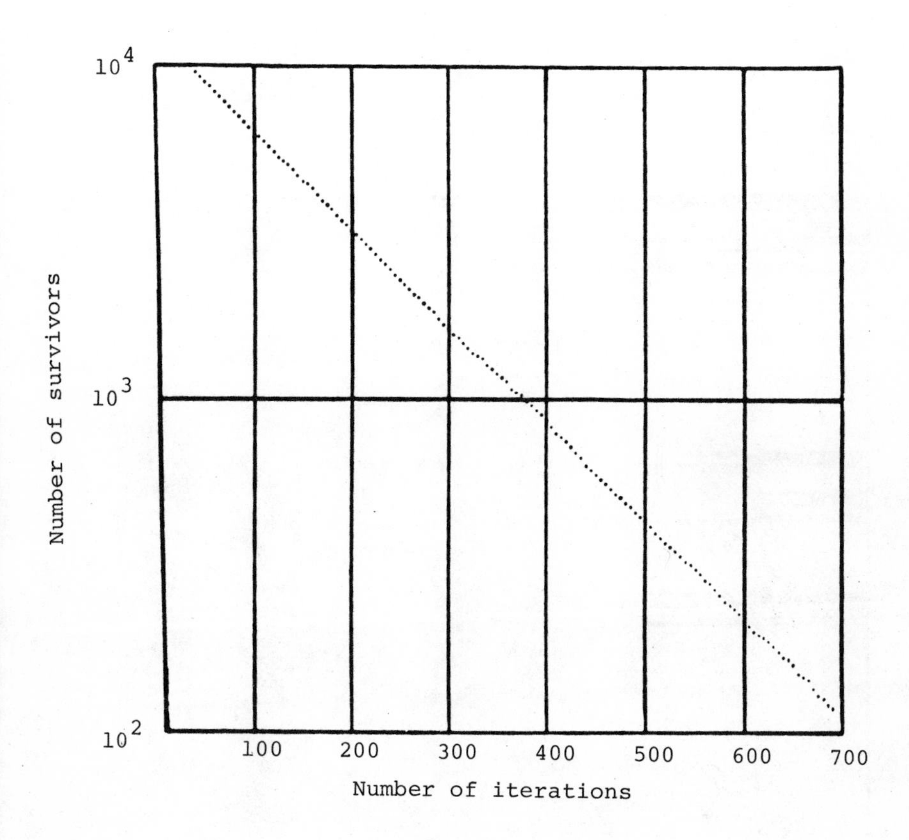

Figure I.16.

fact the orbits of nearby points stay close for almost all choices of the initial points. We have thus an ergodic but not a mixing behavior of the map in this case.

We wish to reformulate the "sensitive dependence" in a more measurable fashion. If we want to know how much we should expect two very close points to separate, we are naturally led to examine the derivative $Df^n = d/dx\ f^n$ of f^n. In the map $f(x) = 1 - 2|x|$ which we have analyzed before, it is obvious from the chain rule of differentiation that $|D\ f^n(x_0)| = 2^n$ unless one of the points $f^k(x_0)$

Fig. I.17. Histogram for 50000 iterates of the point $x = .2$ by the function $f(x) = 1 - 1.401155x^2$; 800 intervals.

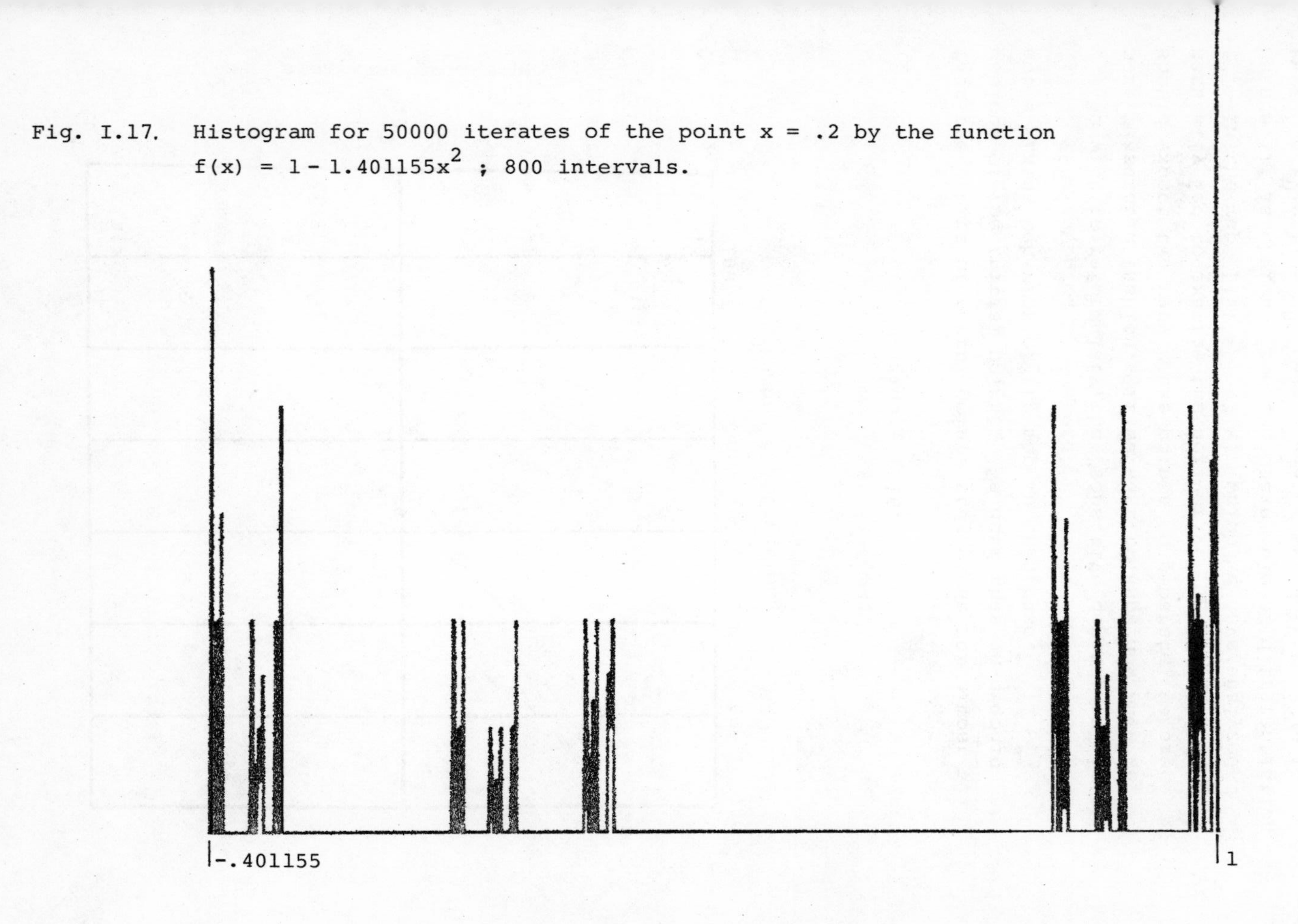

equals zero, and then the quantity $Df^n(x_0)$ is undefined for $n > k$. We see thus that for most initial points the derivative along the orbit diverges like an exponential, namely like $\exp(n \log 2)$. The number $\log 2$ will then be called the characteristic exponent or Liapunov exponent of the map f. Now it so happens that many of the maps which show behavior similar to $x \to 1 - 2x^2$ have positive characteristic exponents (while the maps like $x \to 1 - 1.401155...x^2$ do have zero characteristic exponents). If a map has an invariant ergodic probability density (II.8) like $[\pi(1-y^2)^{1/2}]^{-1}$ in the case of the map $1 - 2x^2$, then for almost all initial points the quantity $\lim_{n\to\infty} 1/n \log|Df^n(x_0)|$ may exist and be positive, say equal to $\alpha > 0$. This will then have the interpretation that in the mean, two initial points which are very close will start to separate at the rate α^n during n iterations. The exact relations between invariant measures, characteristic exponents and sensitive dependence on initial conditions as well as topological and Kolmogorov-Sinai entropy are very subtle and not yet totally clarified, (cf. II.8).

IMPORTANT REMARK. When we have discussed above the "typical" behavior of a map, we have always insisted on analyzing what happens to most initial points. This is motivated by the fact that we want to make general statements about the behavior of dynamical systems. It happens very often that while most of the initial points show a very regular behavior (i.e., they approach a stable periodic orbit), some other initial points--very few in the sense of Lebesgue measure--behave rather in the ergodic way described above. Such a situation has been described in the paper by Li and Yorke [1975] "Period three implies chaos" (cf. II.3). They show, among other things, that if a map has a (stable or unstable) orbit of period three, then the map in question shows sensitive dependence with respect to initial conditions for an uncountable set of pairs of initial points. But--and maybe the paper did not make this sufficiently clear--for <u>most</u> other points this need not be the case, and hence from a physical point of view the chaotic behavior may be essentially unobservable.

implications on the possible outcome of such modelling or experiments. This can be viewed positively or negatively. The negative aspect is that it shows the limitations of modelling itself. The positive aspect is that it gives us indications about a physical system without knowing specific details, and as we shall see, one can predict qualitative and quantitative results for experiments depending on a parameter. We anticipate how a typical result looks. It will say, as an example: "If a certain phenomenon does occur for a given system at some value of the control parameter, then one can expect some specific other phenomenon at some other value of the parameter." Sometimes a prediction can be made for the expected value of the parameter where the second phenomenon is expected to occur.

Most of our considerations will be illustrated for the family of maps $x \to 1-\mu x^2$, on the interval $[-1,1]$, where μ varies in the interval $(0,2]$. The reader may be more accustomed to the well studied family $y \to \gamma y(1-y)$, mapping the interval $[1-\gamma/4, \gamma/4]$ into itself, when $2 < \gamma \le 4$. Under the coordinate change $y = (\gamma/4-\frac{1}{2})x + \frac{1}{2}$, the two families coincide, with the parameter identification $\mu = \gamma^2/4 - \gamma/2$.

We close this section with an outlook on the phenomena we shall discuss in the next sections. For this purpose we perform an "experiment" with the one-parameter family of maps $x \to f_\mu(x)$, where $f_\mu(x) = 1-\mu x^2$. As always, x is the "observable" and μ is the parameter. The experiment, done on a computer, is the following. We fix μ to be one of 800 equally spaced values in the interval $[0.35,2]$. For such fixed μ we take $x_0 = 0$ and iterate the map f_μ. On the vertical axis one point is plotted above μ for each of the numbers $x_{301}^{(\mu)}$ to $x_{425}^{(\mu)}$, where $x_j^{(\mu)} = f_\mu^j(0)$. The purpose of "waiting" 300 iterations until we plot the first point is the following. If f_μ has a stable periodic orbit which is sufficiently attractive, the points $x_{301}^{(\mu)}$ to $x_{425}^{(\mu)}$ will be very near to it. So we can expect the figure to show quite clearly the (sufficiently) stable periodic orbits. Of course we will not be able to distinguish on the figure between a stable periodic orbit with a very long period and between ergodic behavior. The result of the experiment is shown in Figure I.19.

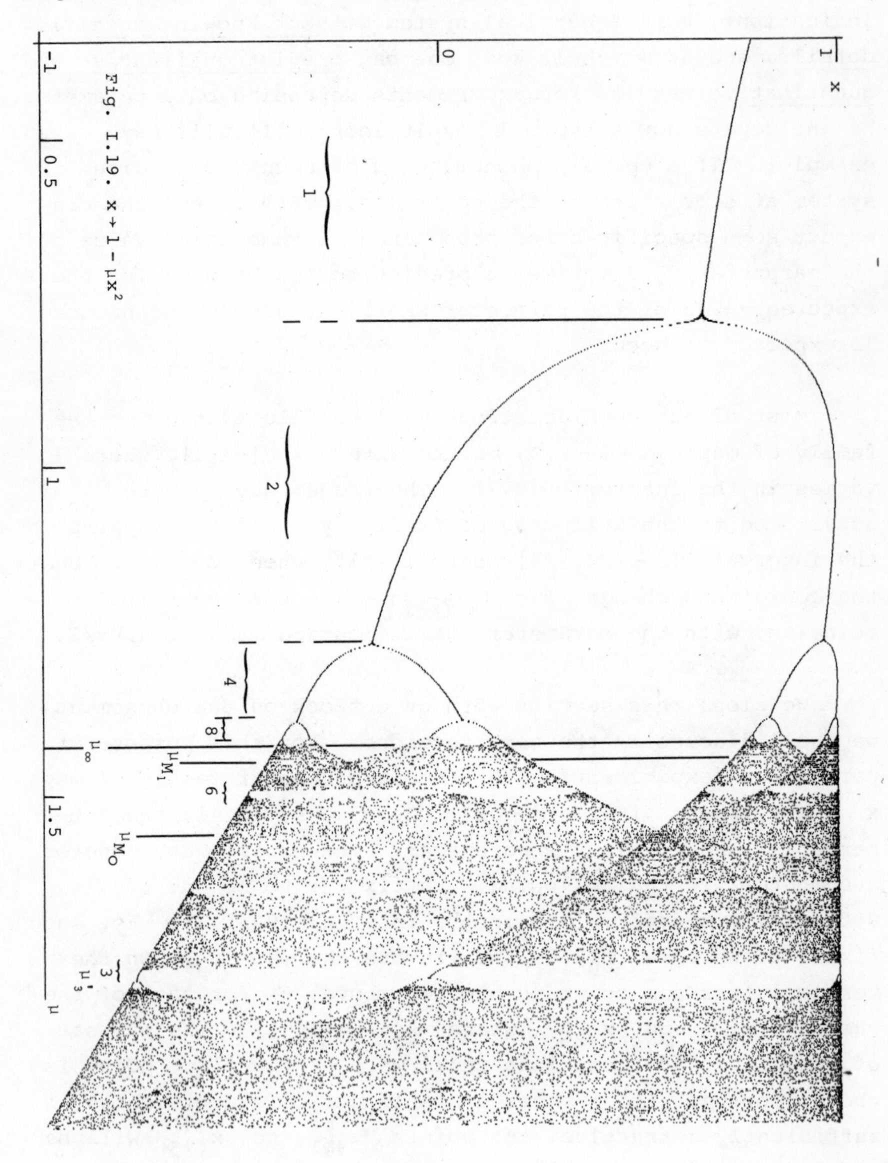

Fig. I.19. $x \to 1 - \mu x^2$

I.4 SYSTEMATICS OF THE STABLE PERIODS

If we look closely at the Figure I.19, we distinguish, from left to right, more or less clearly, a stable orbit of period 1 (i.e., a fixed point), then a stable orbit of period 2, then 4, 8, then a "mess," then another orbit of period 6, then 5, and 3. Other stable periods are barely visible in-between. The astonishing fact about this arrangement of stable periodic orbits is its independence of the particular one-parameter family of maps. A complete description of this arrangement will be given in III.1, based on earlier developments in II.2. We describe here some of its aspects. Note first of all, that there are different kinds of stable periods of the same length. This differentiation comes from the order in which the points are visited. As an example, there are two different stable periods of length four (the second one is barely visible to the right of the period 3 in Fig. I.19).

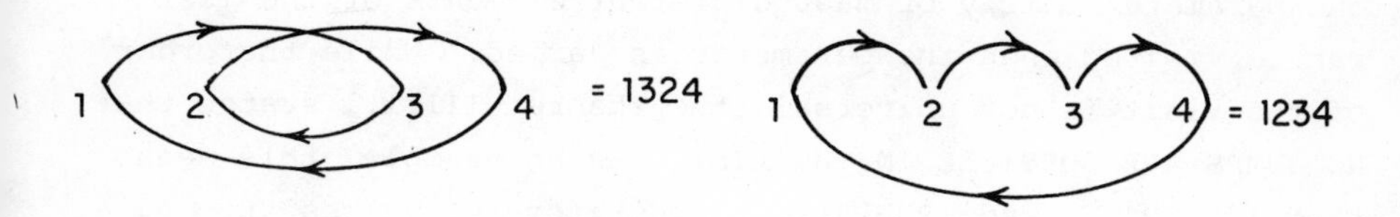

Figure I.20

(The possibilities

do not occur for the families we are considering, since in our case the image of the rightmost point must always be the leftmost point.) Note that the two cases can be distinguished in an experiment, since they have different power spectra.

One can give a complete listing of all the possible periodic orbits of some bounded length. We give here a list for periods of length up to 6, in the sense of Fig. I.20.

TABLE I.21

PERIOD	APPEARS FIRST AT	ORDER OF POINTS	TYPE
2	-.25	12	RR
4	.75	1324	RLRL
6	1.475	143526	RLRRRL
5	1.624	13425	RLRRR
3	1.750	123	RLR
6	1.769	135246	RLLRLL
5	1.861	12435	RLLRL
6	1.907	124536	RLLRRR
4	1.941	1235	RLLR
6	1.967	123546	RLLLRL
5	1.985	12345	RLLLR
6	1.996	123456	RLLLLR

The interpretation of this list is as follows. In a continuous one-parameter family of maps different elements of the list can be visited when the parameter is varied. While the order of the visit is not universal, the Theorem III.1.1 states that NO JUMPS ARE POSSIBLE IN THE LIST. As an example, this means that if RR... and RLRRRL... occur for two values $\mu_1 < \mu_2$ of the parameter, then RLRL <u>must</u> occur for a value μ_3 which lies <u>between</u> μ_1 and μ_2, i.e., $\mu_1 < \mu_3 < \mu_2$. Let us stress, however, that this theorem is only true for once differentiable maps of the interval into itself, and that it does not readily generalize to maps on $\mathbb{R}^n$ or to flows.

As a preliminary remark, we note: In the list, after each period of length k there follows immediately an infinite sequence of periods $k\cdot 2$, $k\cdot 4$, $k\cdot 8,\ldots,k\cdot 2^n,\ldots$. We see this quite clearly for the case $k = 1$, in Fig. I.19 and less clearly for $k = 3$. In particular, there is nothing else (i.e., no aperiodic region) between $k\cdot 2^n$ and $k\cdot 2^{n+1}$. We shall see later that these sequences of "subharmonic bifurcations" exhibit strong numerical relations for the parameter values. But before analyzing this question, we shall investigate in the next section the relative frequency of periodic and aperiodic behavior.

I.5 ON THE RELATIVE FREQUENCY OF PERIODIC AND APERIODIC BEHAVIOR

In this section, we temporarily relax the principle of describing only proved theorems and not any conjectures. The question addressed in the title is a very delicate one and it has only been answered for specific, isolated one-parameter families of maps. Our description will thus be inspired by the experience one has collected in proving these isolated facts.

Let us look first at the stable periods. As we have said before, $\bar{x}$ is on a stable period of (primitive) length k if

1) $f_\mu^j(\bar{x}) \neq \bar{x}$ for $j = 1,\ldots,k-1$,

2) $f_\mu^k(\bar{x}) = \bar{x}$,

3) $|f_\mu^{k\,\prime}(\bar{x})| < 1$.

These three conditions show us immediately that if f_μ has a stable period of length k, and if f_μ depends sufficiently regularly on μ, then for nearby μ', $f_{\mu'}$ will also have a stable periodic orbit, by continuity. Of course $\bar{x}$ itself varies with μ. May [1976] has introduced the term <u>window</u> for the (open) interval in parameter space in which the period k remains stable. Let us redraw the windows corresponding to the shortest periods (of Fig. I.19).

Figure I.22

Note that the windows tend to get shorter as the periods get longer.

The problem we consider is now the following. Take the parameter interval and discard all windows corresponding to all stable periods. What remains?

Let us illustrate why it is possible that something remains although an infinite set of windows has to be discarded. If, in the interval $(0,1)$, we enumerate the rationals and if around the n-th rational we discard an interval of length $\varepsilon 2^{-n}$, the volume (i.e., the Lebesgue measure) of what is left over is at least

$$1 - \sum_{n=1}^{\infty} \varepsilon 2^{-n} = 1 - \varepsilon .$$

Nevertheless the left-over set contains no intervals.

This mathematical example illustrates what can happen. Imagine that each rational number corresponds to a value of the parameter μ for which we find a stable period, and each small interval around this value corresponds to the window described above. Then the left-over region will correspond to values of the parameter for which there is no stable periodic orbit. We conjecture :

1) The set of values of the parameter for which there is no stable periodic orbit has positive Lebesgue measure.

2) It contains no intervals.

This says that if we choose a parameter μ "at random" (in the Lebesgue sense), there is a non-zero probability to find an aperiodic behavior for the map.

We describe next somewhat more precisely the relative frequency of aperiodic behavior. For this we specialize to the family $f_\mu(x) = 1 - \mu x^2$. (The theorems are proven for a slightly different family). We have already argued that when $\mu = 2$, the map f_μ can be brought by a coordinate transformation to the form $g(x) = 1 - 2|x|$. Whenever, in a family of transformations, such a change of coordinates is possible, (or some similar change of coordinates is possible) then we conjecture the following result (which has been proved in particular instances (III.2)).

Consider the interval $\mu \in [2-\varepsilon, 2]$. Denote by V_ε the volume (Lebesgue measure) of the set of those parameters for which aperiodic behavior occurs. We compare this volume to the volume ε of the interval $[2-\varepsilon, 2]$. Then we have

$$\lim_{\varepsilon \to 0} V_\varepsilon/\varepsilon = 1 \qquad ,$$

i.e., as ε gets smaller, <u>most</u> of the values of the parameter lead to <u>aperiodic</u> behavior.

A nice computer experiment has been performed by Shaw [1978] to illustrate this. He measures the aperiodicity by looking at the characteristic exponent, as a function of μ, for the family $x \to \hat{f}_\mu(x) = \mu x(1-x)$. What has actually been computed is

$$\bar{\lambda}(\mu) = \frac{1}{100000} \sum_{j=1}^{100000} \log|\hat{f}'_\mu(\hat{f}^{j-1}_\mu(0))|/\log 2$$

for 300 values of μ spaced .002 apart. See Figure I.23.

We see that the "windows" lead to negative $\bar{\lambda}$, while aperiodic behavior seems to lead to $\bar{\lambda} > 0$. The nearer we get to $\mu = 2$, the fewer "dips" become observable. In part this is due to the fact that they get very thin. One can show

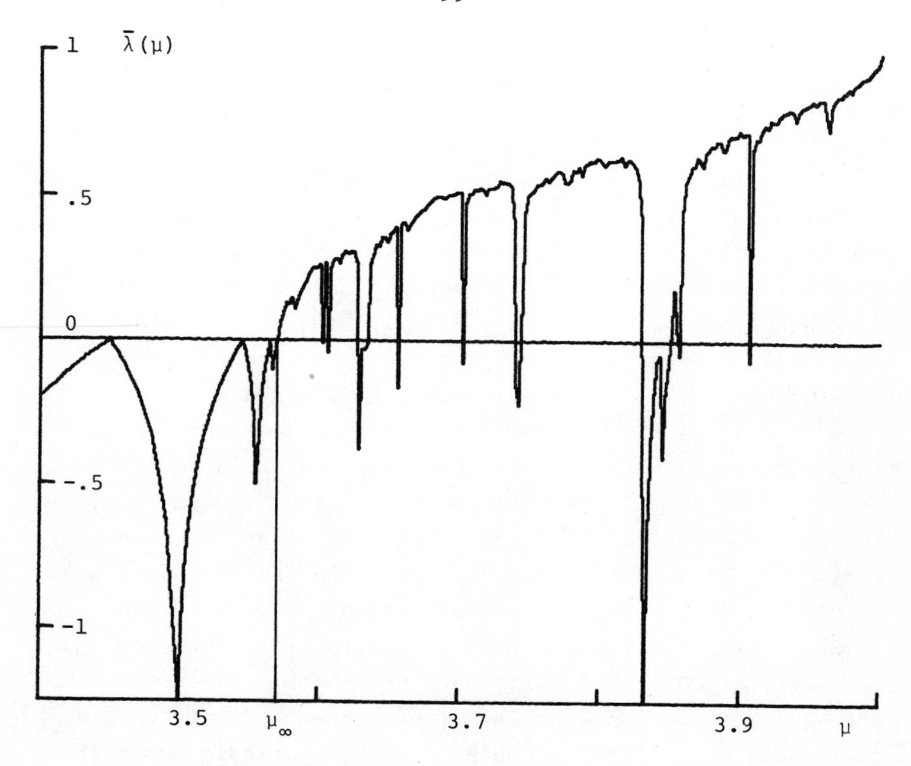

Figure I.23.

(outside the set of parameters leading to periodic behavior) that $\bar{\lambda}(\mu)$ behaves as $1 - \text{const.}(4-\mu)^{1/2}$ and this compares well with the figure.

We also note that a similar picture has been obtained by Feit [1978] for the Hénon maps, but this is even less understood.

Let us now analyze in more detail the behavior of f_μ for μ in the aperiodic region. The situation is as follows. It is believed that the set V^1 of those μ for which f_μ has no stable periodic orbit has positive Lebesgue measure, as we have said before. This set contains a smaller subset V^2, still of <u>positive Lebesgue measure where</u> f_μ has <u>sensitive</u>

dependence on initial conditions. Things get now slightly more complicated.

It is rigorously known that the set V^3 of μ for which f_μ has an absolutely continuous invariant ergodic measure is uncountable (cf. Pianigiani [1979], Misiurewicz [1977], III.2). A function with this property will lead to a density (histogram) of the points x_0, $f(x_0)$, $f^2(x_0)$,... which is essentially independent of x_0. It has been announced by Jakobson [1978,1979] that even the set V^3 has positive Lebesgue measure, but no detailed proof is available, yet. We tend to believe the result, however. Let us describe in somewhat more detail how the uncountable set discovered by Misiurewicz looks (II.8). The first idea is a direct generalization of the situation found by Ulam and von Neumann [1947] and exploited by several authors; Ruelle [1978(1)], Bowen [1975]. It is the situation which occurs if an image of the maximum (i.e., of zero) falls onto an unstable fixed point, or an unstable periodic orbit. The argument of Section I.2 shows again that there cannot be a stable periodic orbit in this case. It is then intuitively easy to understand why there is sensitive dependence on initial conditions: As we have already seen, sensitive dependence means that nearby points are eventually "pulled apart" through the iteration. This condition is hardest to fulfill for points near $x = 0$, since the derivative of f_μ at zero is zero, and hence f_μ contracts points near zero. But eventually zero itself is mapped onto an unstable fixed point (or periodic orbit). That is, it stands still. But a point near to zero will not be mapped onto the fixed point (or periodic orbit), and hence it will be pushed away by further iterations from the fixed point (or periodic orbit) because the latter is unstable. This is illustrated in Figure I.24.

Misiurewicz has in fact shown more, namely that the above conclusions hold provided there is a neighborhood of zero which is not visited by any image $f_\mu^n(0)$ of zero. This leads to a very large class of values of the parameter, but it is at

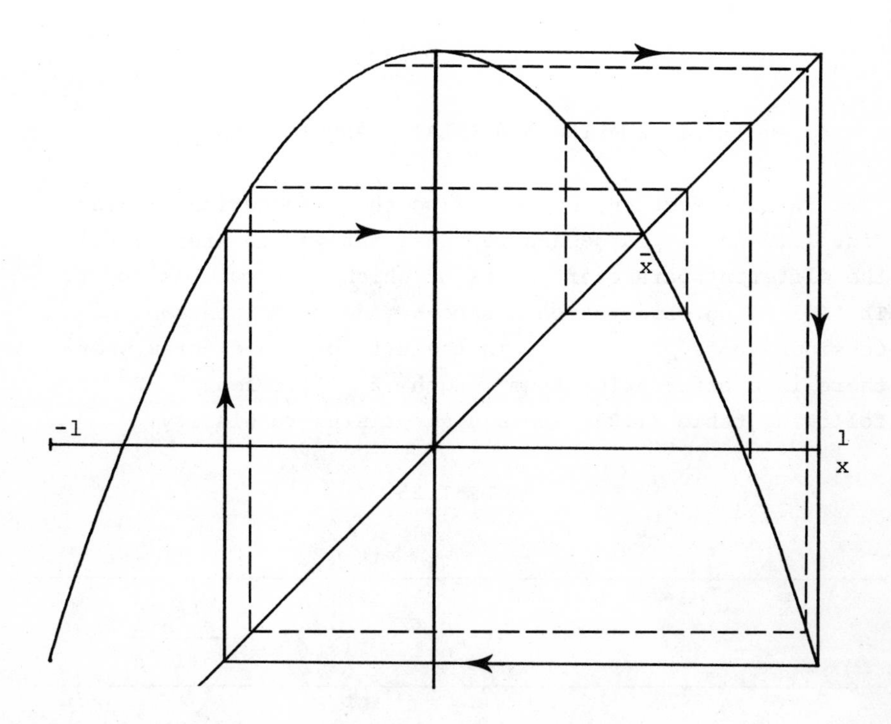

Figure I.24. $f(x) = 1 - 1.544x^2$

least intuitively clear (III.2) that the Lebesgue measure of such μ must be zero. Therefore there are "many" other values of μ for which f_μ has no stable periodic orbit, and sensitivity on initial conditions, but for which we do not know whether or not f_μ has an absolutely continuous invariant measure.

I.6 SCALING AND RELATED PREDICTIONS

In this section, we reexamine the bifurcation diagram Fig. I.19 in the large and we shall analyze in great detail the different parameter values at which bifurcations occur. The following observation has been made by Feigenbaum. Consider those points μ_n in the bifurcation diagram where there is a bifurcation from length 2^{n-1} to length 2^n. The following table (I.25) shows a surprising regularity.

TABLE I.25

n	μ_n	$\mu_n - \mu_{n-1}$	$\dfrac{\mu_n - \mu_{n-1}}{\mu_{n+1} - \mu_n}$
0	.75		
		.5	
1	1.25		4.233738275
		.1180989394	
2	1.3680989394		4.551506949
		.0259472172	
3	1.3940461566		4.645807493
		.0055850823	
4	1.3996312389		4.663938185
		.0011975035	
5	1.4008287424		4.668103672
		.0002565289	
6	1.4010852713		4.668966942
		.000054943399	
7	1.401140214699		4.669147462
		.000011767330	
8	1.401151982029		4.669190003
		.000002520208	
9	1.401154502237		4.669196223
		.000000539752	
10	1.401155041989		

1) We see that the μ_n seem to converge to some number μ_∞.

2) The ratios $(\mu_n - \mu_{n+1})/(\mu_{n+1} - \mu_{n+2})$ seem to converge to some number $\delta = 4.66920...$.

THEOREM. (Feigenbaum [1978], Collet-Eckmann-Lanford [1980]). <u>For sufficiently smooth families of maps of the interval to itself the number</u> δ <u>does not in general depend on the family</u> (III.3).

Let us expand a little bit on this. The statement says that for any "nice" one-parameter family of maps, if we are not very unlucky in that the parameter dependence is very special near μ_∞ then <u>asymptotically</u>

$$|\mu_n - \mu_\infty| \sim \text{const. } \delta^{-n} \quad ,$$

where δ is universal.

This sort of result is of course strongly reminiscent of the renormalization group analysis and in fact such that a group is lurking behind this. We also observe that the limit in (2) is approached quite quickly.

All this can be reformulated in the following way. If we choose as a new coordinate the quantity $\log|\mu - \mu_\infty|$ then the bifurcation diagram should have bifurcations at asymptotically regular intervals. This is well illustrated in Figure I.26. We can see clearly how the asymptotic regime is reached. We next analyze the behavior in the variable x. Again there is some universal behavior.

THEOREM. <u>Some geometric feature scales asymptotically like</u> λ^n <u>at the n-th bifurcation, where</u> λ <u>is a universal constant</u>, $\lambda = -0.3995...$.

To illustrate this, we may plot, for each value μ_n' for which 0 is on a stable period of length 2^n, the 2^n points of this orbit. Such periods are called superstable, since $|f^{2^n\prime}(0)| = 0$ which is the strongest linear contraction we

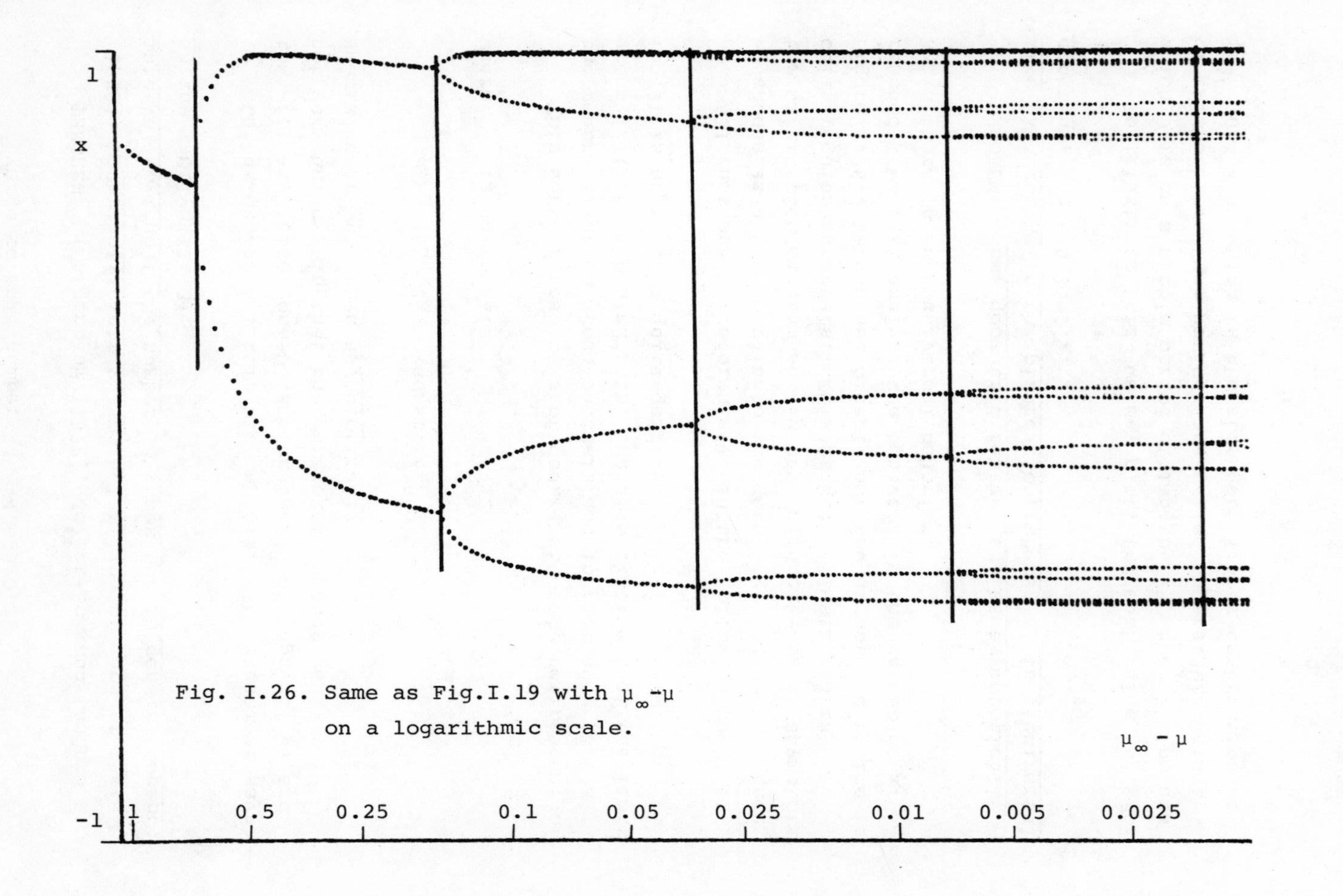

Fig. I.26. Same as Fig.I.19 with $\mu_\infty - \mu$ on a logarithmic scale.

can encounter for a given period. The outcome of the calculation is given in Fig. I.27. We see that the point nearest to zero has distance about λ^n from zero. A more striking way to illustrate this regularity is to rescale the x axis in a μ-dependent fashion. If the horizontal axis is $\log|\mu-\mu_\infty|$ as before and if the scale on the vertical axis is $x\cdot|\mu_\infty-\mu|^{-0.595367\ldots}$ then we will find asymptotically a periodic diagram, with reflections (the number $\alpha = -0.595367\ldots$ is given by the equation $\delta^\alpha = |\lambda|$), see Figure I.28.

Recall that all these statements and numbers are <u>independent</u> of the one parameter family in question. Figure I.29 shows this independence for a multitude of cases.

We have plotted $\log|\mu_n-\mu_\infty|$ versus n, where μ_n is defined to be the point of bifurcation for the models of Hénon (Derrida-Gervois-Pomeau [1979]), of Lorenz (Franceschini [1979]) and of Navier-Stokes on a torus (Franceschini-Tebaldi [1979]), while it is the point of superstable periods for the maps $x \to \mu(1-2x^2)$ and $\mu x(1-x^2)$. A similar universality can be seen for the scaling in the x-direction (for the Hénon map we take the Euclidean distance, $\bar{x}$ is zero for maps on the interval), see Figure I.30. We also stress that the phenomenon is not only mathematically understood, but additionally it is very robust in the following sense: If a one-parameter family shows subharmonic bifurcations to periods 1,2,4, for example, then there is so to speak a higher probability of finding a period 8 at a further parameter variation than of finding a period 4 when only periods 1 and 2 have appeared.

We shall leave for the moment the map f_{μ_∞} aside, and we shall concentrate on the "other" side of μ_∞ of Fig. I.19. The theory predicts the same scaling behavior for the other side, and we can see this quite clearly in the next two figures, which represent the scaling in the μ-direction above and then in x and μ, respectively, as before. See Figures I.31 and I.32.

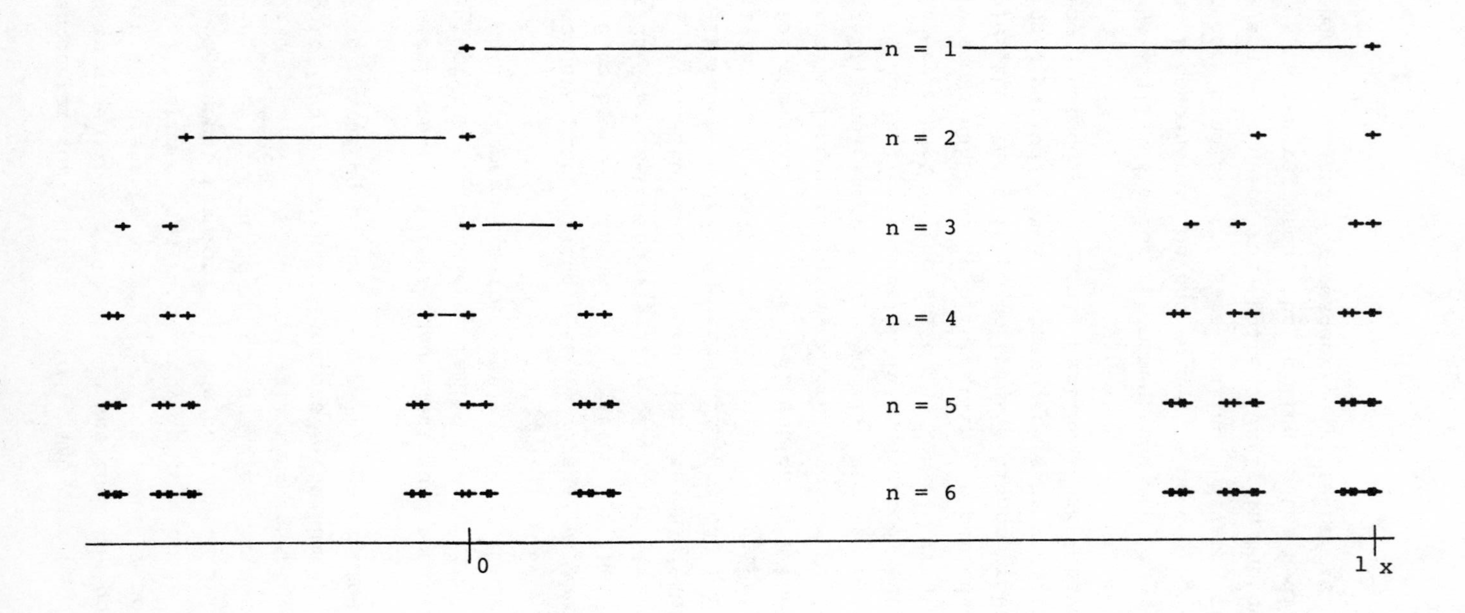

Fig. I.27. Each horizontal line represents the 2^n points of a superstable periodic orbit for a map of the form $x \to 1 - \mu x^2$.

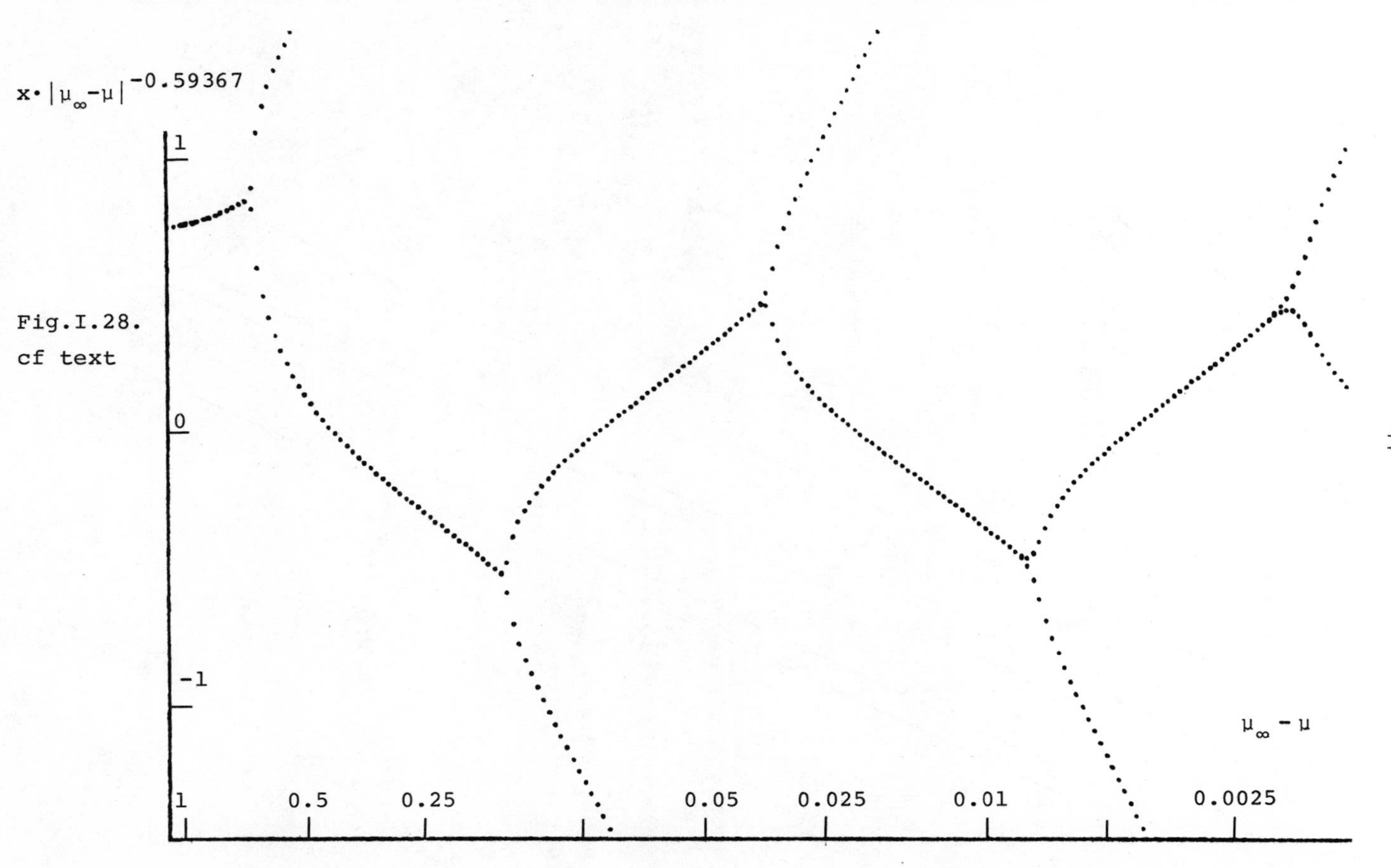

Fig.I.28.
cf text

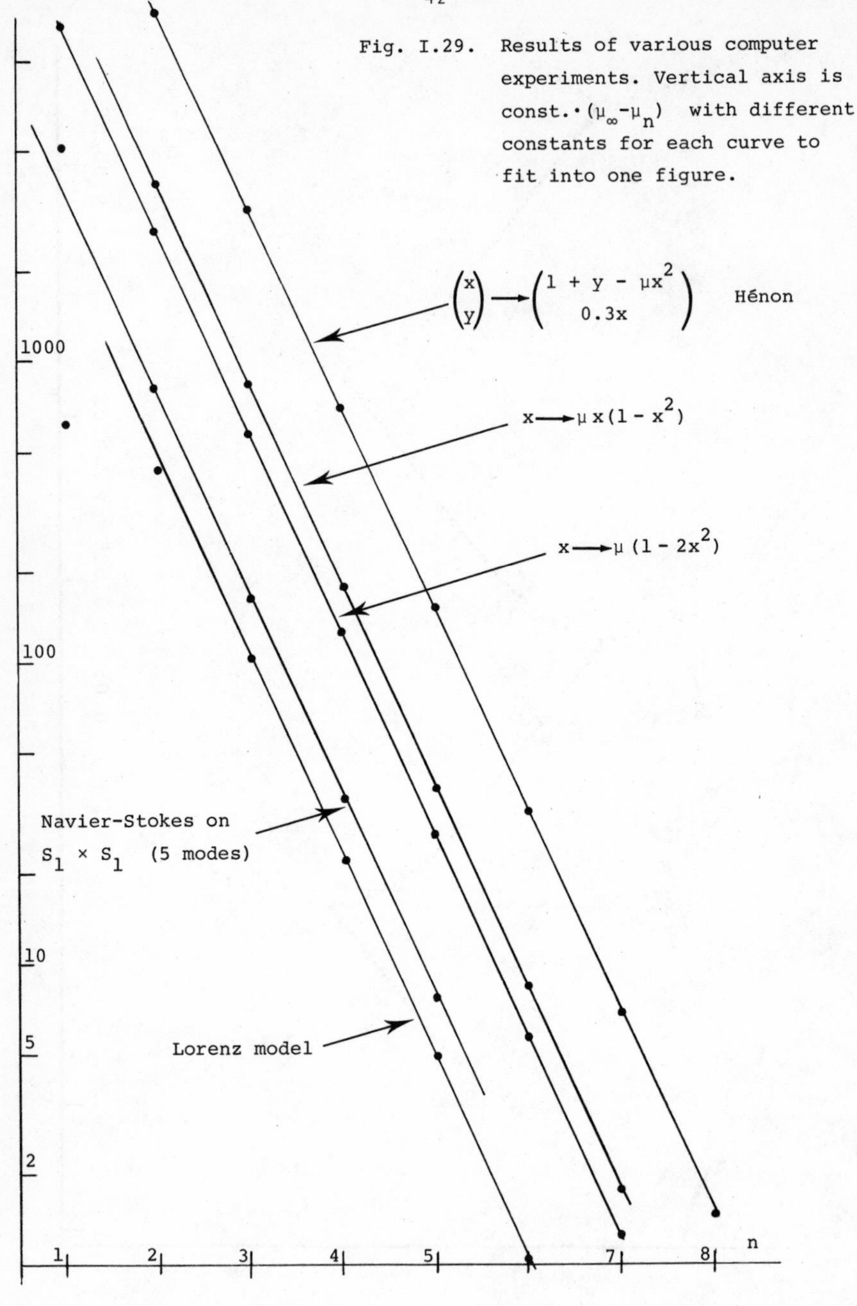

Fig. I.29. Results of various computer experiments. Vertical axis is const.$\cdot(\mu_\infty - \mu_n)$ with different constants for each curve to fit into one figure.

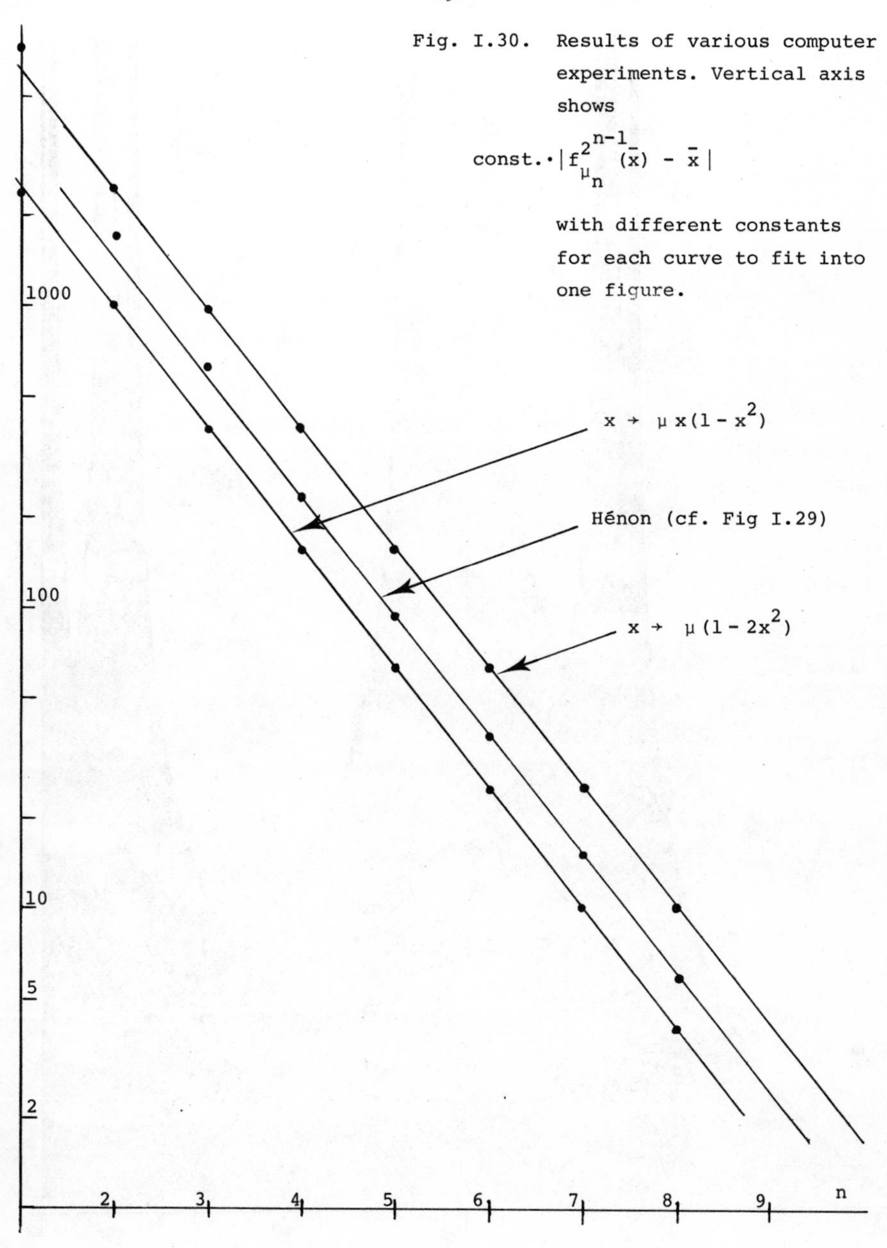

Fig. I.30. Results of various computer experiments. Vertical axis shows

$$\text{const.}\cdot\left|f_{\mu_n}^{2^{n-1}}(\bar{x}) - \bar{x}\right|$$

with different constants for each curve to fit into one figure.

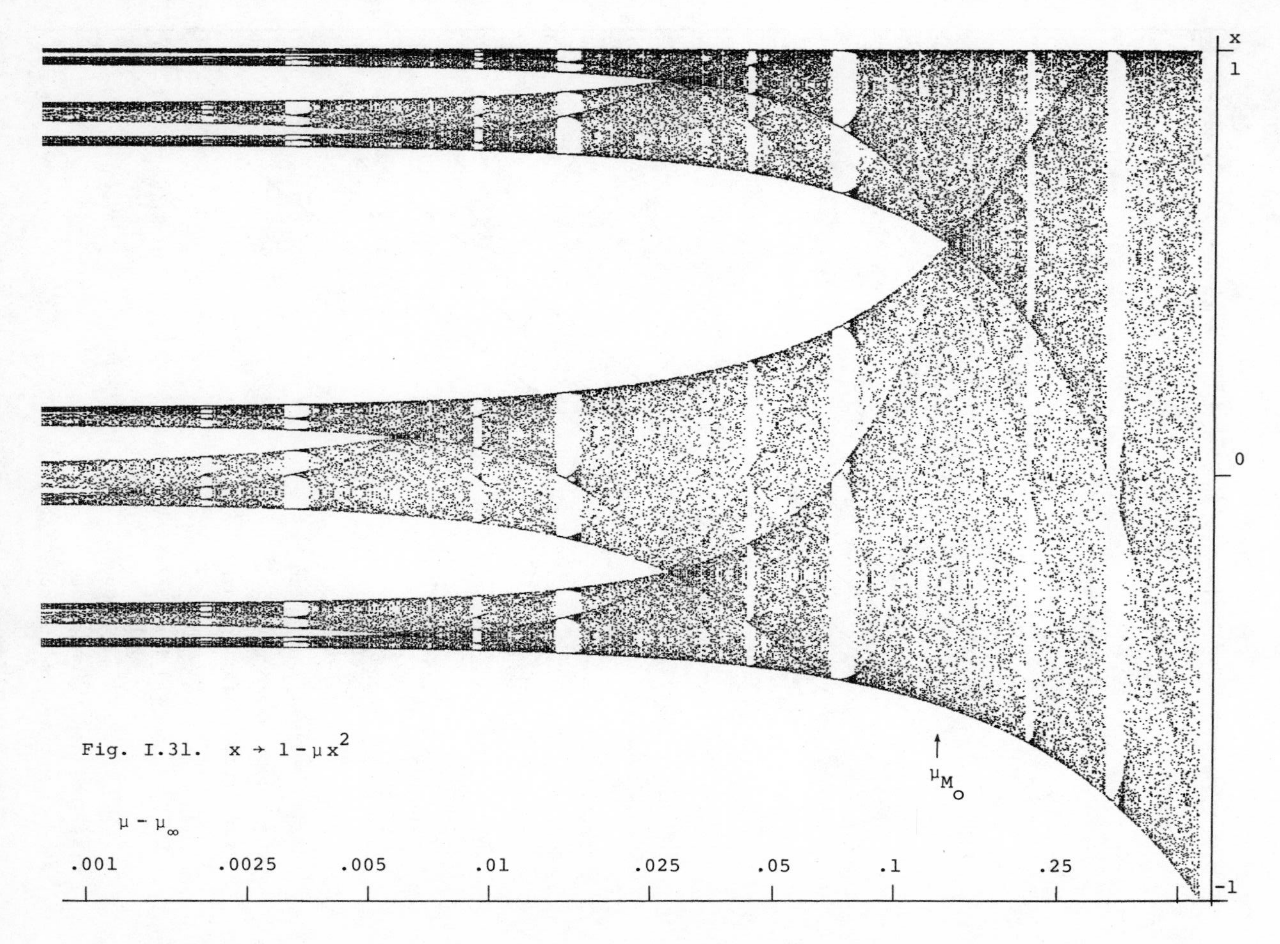

Fig. I.31. $x \to 1-\mu x^2$

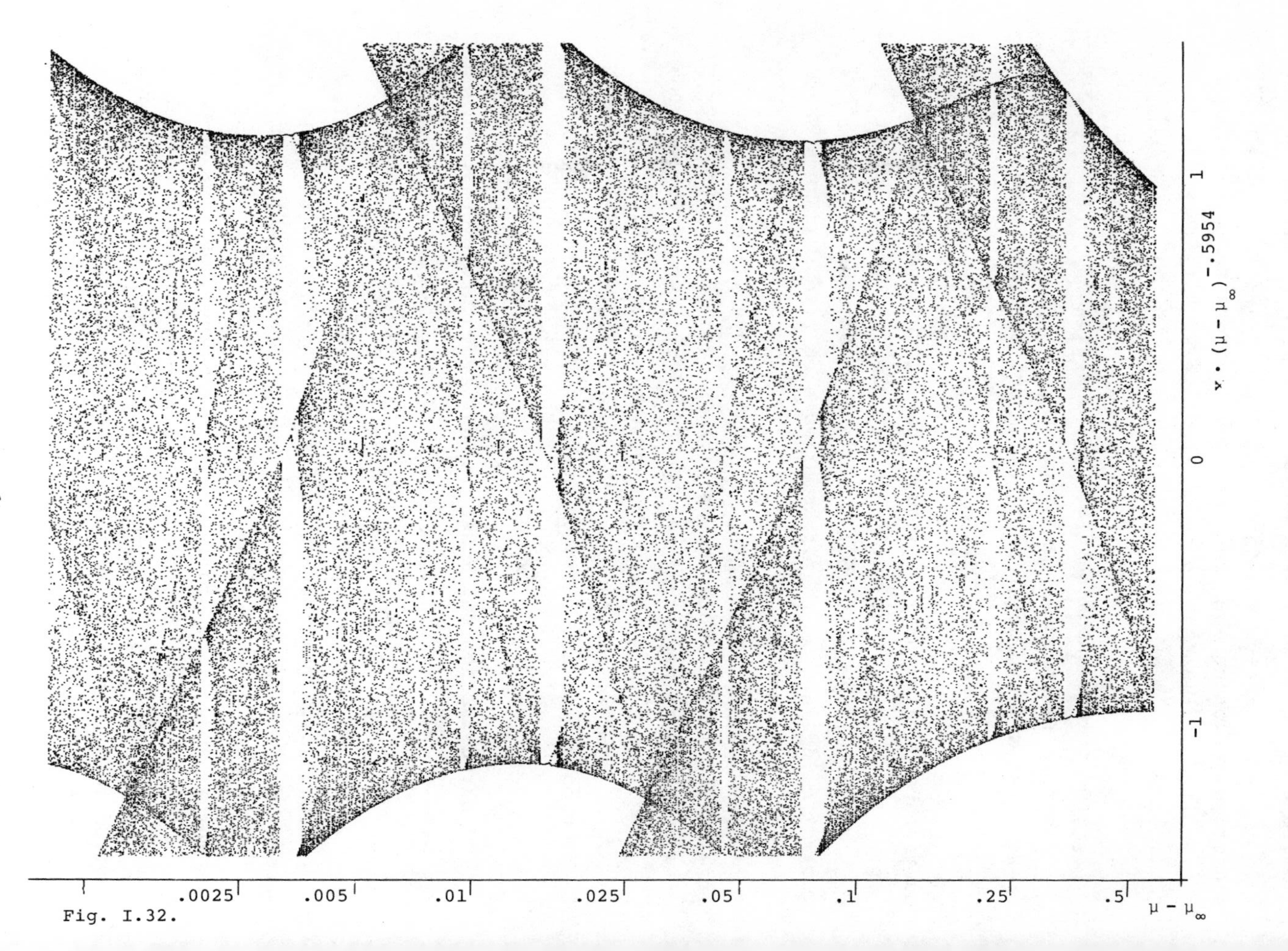

Fig. I.32.

We observe that all features are repeated in a regular fashion, by superposition of a large periodic motion of period 2^k, with a periodic or aperiodic motion on the small scale. Recall now the examples from the preceding section of how to produce very aperiodic maps by insisting, e.g., that $f_\mu^3(0)$ falls on an unstable fixed point. This value is marked μ_{M_0} in the Figure I.31 and in Fig. I.19, and it has the histogram A given below. Now, according to the general theory there will be at the appropriately chosen value of the parameter $\mu_{M_1} \sim (\mu_{M_0}-\mu_\infty)/\delta + \mu_\infty$ a map such that $f^6_{\mu_{M_1}}(0)$ falls on an unstable period 2. At $\mu_{M_2} \sim (\mu_{M_0}-\mu_\infty)/\delta^2 + \mu_\infty$, $f^{3\cdot 2^2}_{\mu_{M_2}}(0)$ falls on an unstable period 4, etc. We show the corresponding histograms in Fig. I.33.B and C.

To summarize, we have found so far a scaling on both sides of the critical value μ_∞. Let us analyze now the neighborhood of the stable period 3 in Fig. I.19. By considering a new family $g_\mu = f_\mu^3$ we are faced with a stable fixed point of g_μ at $\mu = \mu_3'$. By the same analysis as before it can (and will) happen that g_μ will bifurcate to periods 2,4,8,..., near points $(\mu_3'-\mu_\infty')/\delta^k + \mu_\infty'$ for some μ_∞'. This means that f_μ will have periods $3\cdot 2^{k-1}$ at $\mu \sim (\mu_3'-\mu_\infty')/\delta^k + \mu_\infty'$. We see this sequence in the righthand portion after μ_3' in Fig. I.19. A magnification with scale $\log|\mu-\mu_\infty'|$ will give again a spacing of $\log \delta$ between successive bifurcations. See Figure I.34.

We conclude that the whole bifurcation diagram is criss-crossed by an infinity of asymptotically universal ratios, independently of whether we are talking about stable periods or aperiodic behavior.

The analysis has actually been carried out for period tripling and quadrupling, and one finds new universal constants, e.g., for tripling $\delta_3 = 55.247...$, $\lambda_3 = -0.107789$. For quadrupling there will be a new universal sequence in addition to "twice doubling", corresponding to the second period 4 of

Fig. I.33A. Histogram for 20000 iterates of the point $x = .2$ by the function $f(x) = 1 - 1.5436x^2$ in 800 intervals

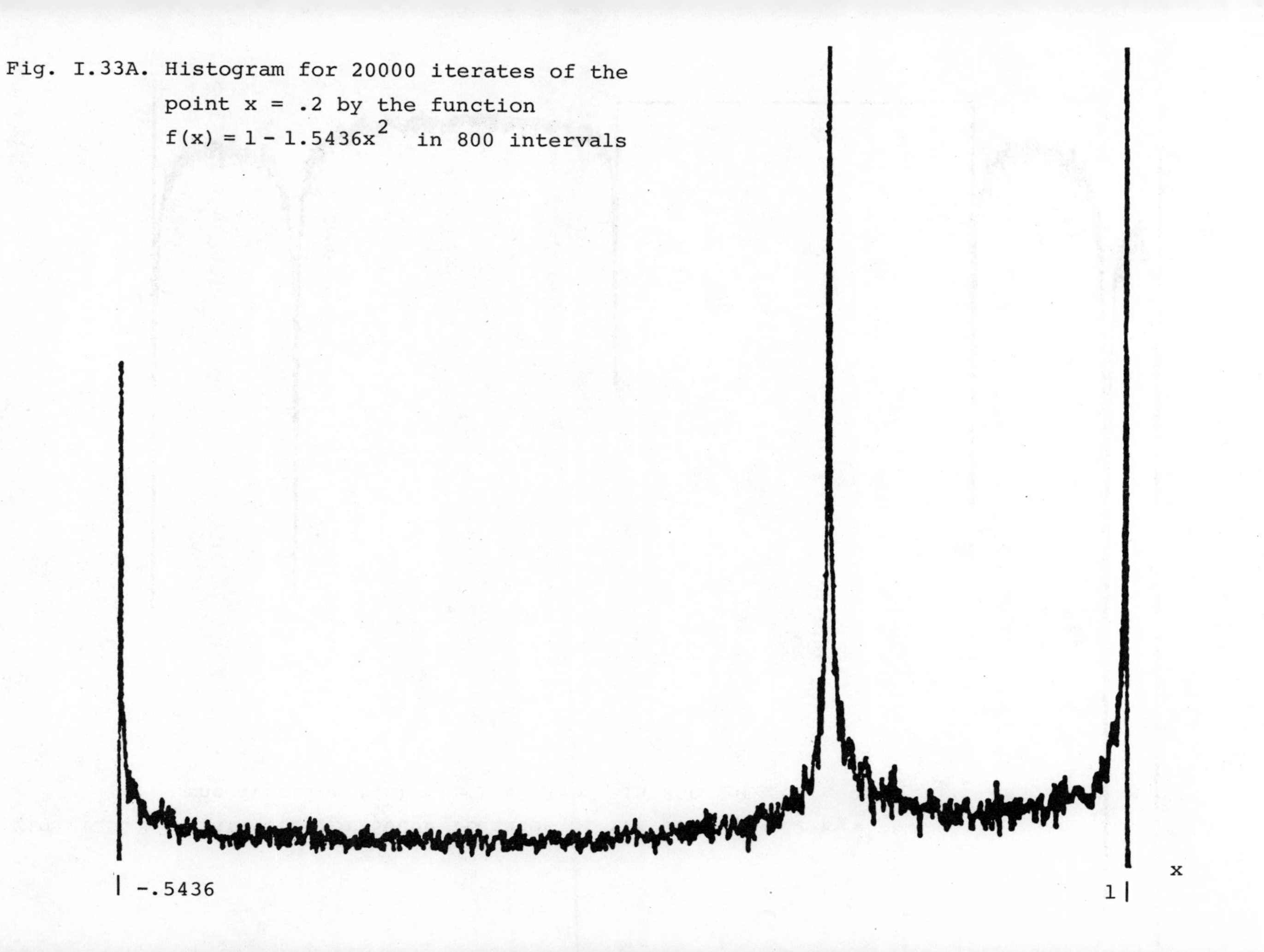

Fig. I.33 B. Histogram for 50000 iterates of the point $x = .2$ by the function $f(x) = 1 - 1.43035\,x^2$ in 800 intervals

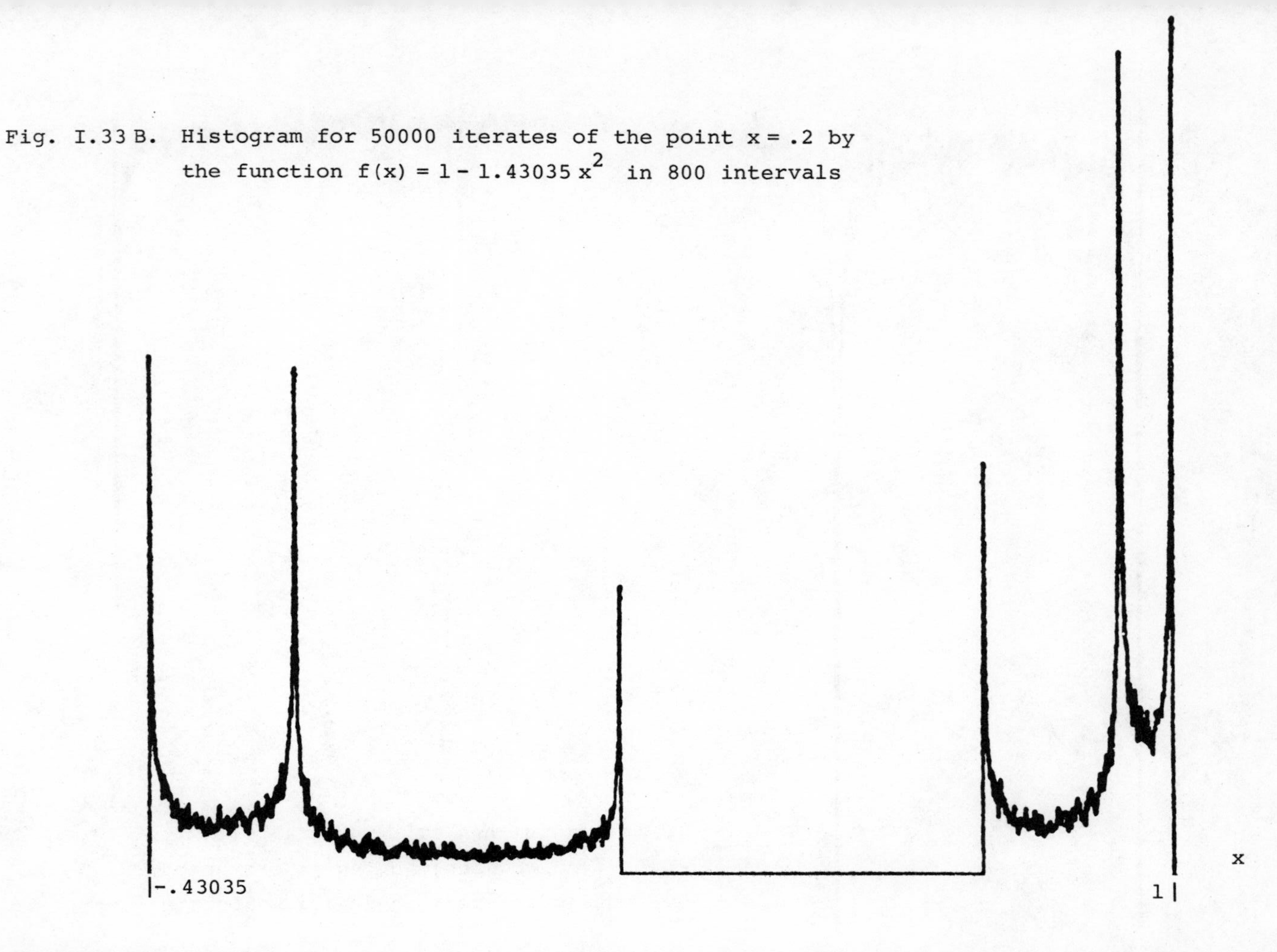

Fig. I.33C. Histogram of 100000 iterates of the point $x = .2$ by the function $f(x) = 1 - 1.40741\,x^2$ in 800 intervals

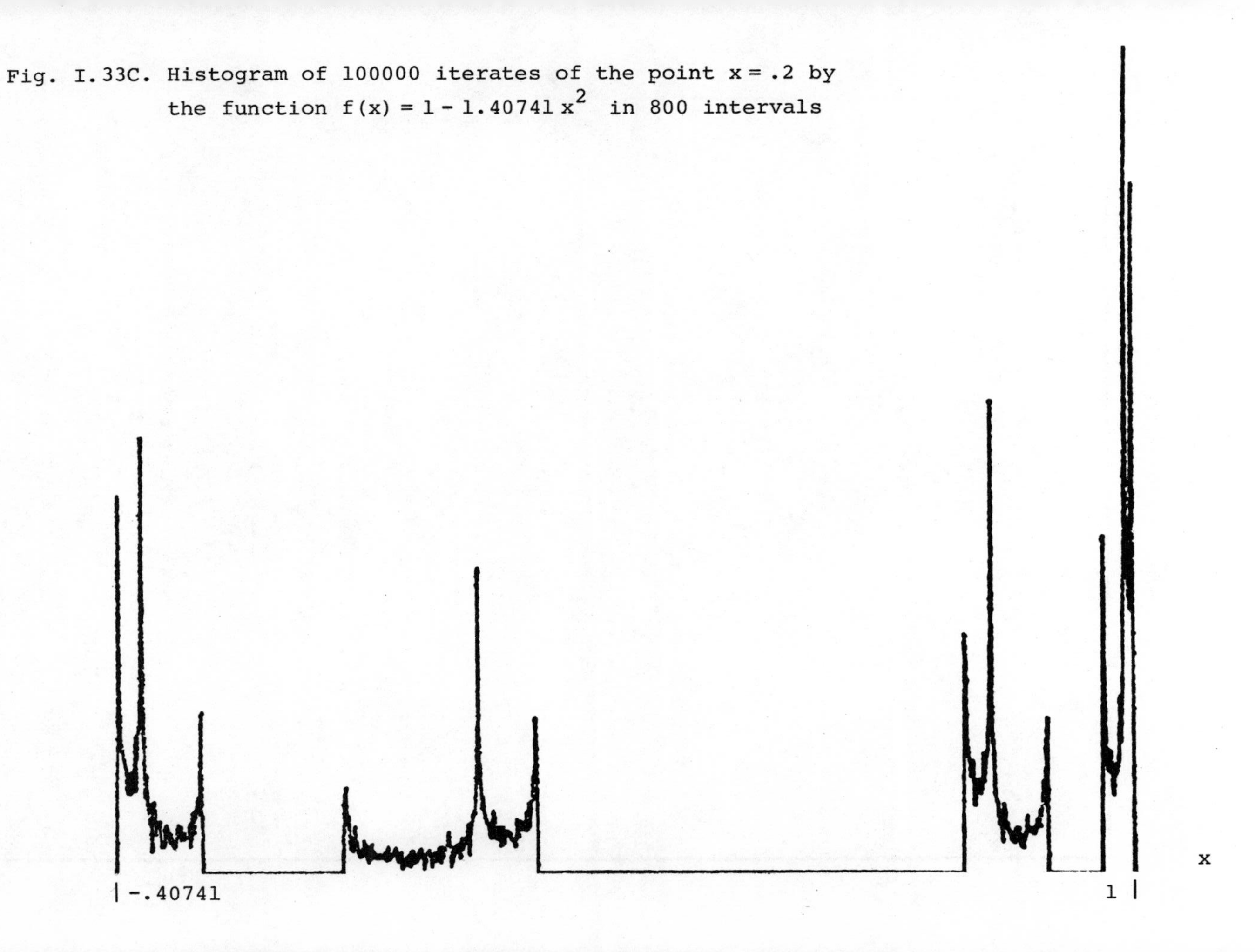

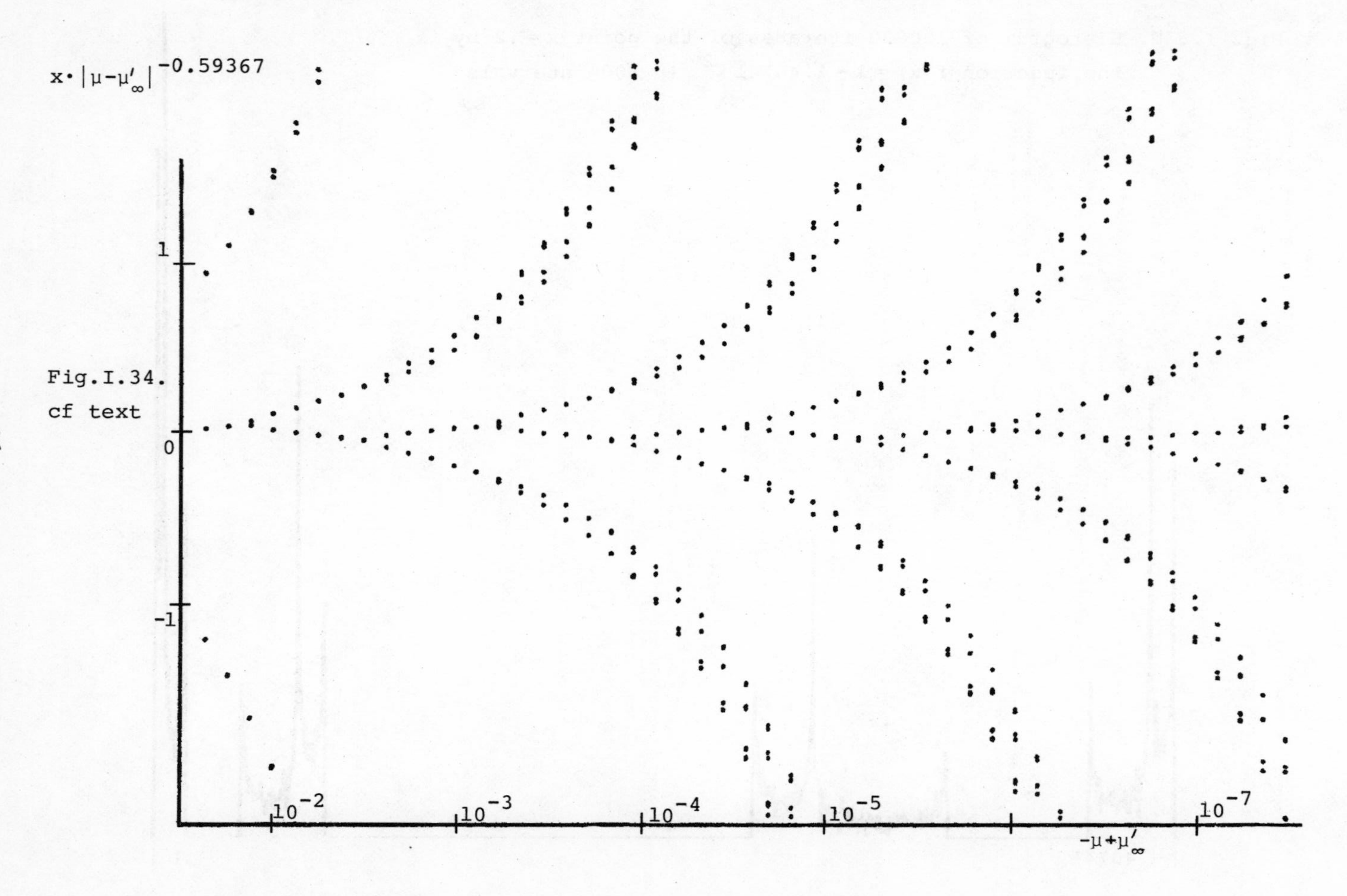

Fig.I.34.
cf text

Fig. I.19. See Derrida-Gervois-Pomeau [1979] for more details about this.

Let us close this section with some general predictions. We have so far only considered one property at a time, either some period $k \cdot 2^n$ or some aperiodic behavior. But in fact there are, asymptotically, universal ratios for the occurrence of different behavior. We just give an example. If some system (physical or numerical) shows two successive period doubling bifurcations for values μ_0 and μ_1 of a parameter, then one may expect other bifurcations near $\mu_j = \mu_0(\delta^{1-j} - 1)/(\delta-1) + \mu_1(\delta-\delta^{1-j})/(\delta-1)$ and "chaotic behavior" near $\hat{\mu}_j \sim (\delta\mu_1-\mu_0)/(\delta-1) - \delta^{1-j}(\mu_0-\mu_1)0,33241/\delta-1$ and period $3 \cdot 2^j$ near $\mu_j' \sim (\delta\mu_1-\mu_0)/(\delta-1) - \delta^{1-j}(\mu_0-\mu_1)0.803/\delta-1$ (this is the first sequence described above), etc. If one wishes to find new such relations, one just reads them numerically off the bifurcation diagram of some simple family (e.g., $x \to 1 - \mu x^2$). They will then hold asymptotically for any other family f_μ.

Very often in physics one does not measure points in phase space, but rather power spectra, i.e., the Fourier transform

$$a_k = \lim_{n\to\infty} \frac{1}{n} \sum_{\ell=1}^{n} e^{2\pi i k \ell} f^\ell(x_0).$$

See Figures I.35 and I.36.

Fig. I.35. Experimental power spectra for the Bénard experiment for Rayleigh numbers of $40.5\ R_c$ and $42.7\ R_c$ (next page), and $43\ R_c$ (top of page 53). Vertical scale is logarithm of amplitude in dB .

Fig. I.36. Power spectrum of the superstable period of length 64 of a map near $f(x) = 1 - 1.401155x^2$.

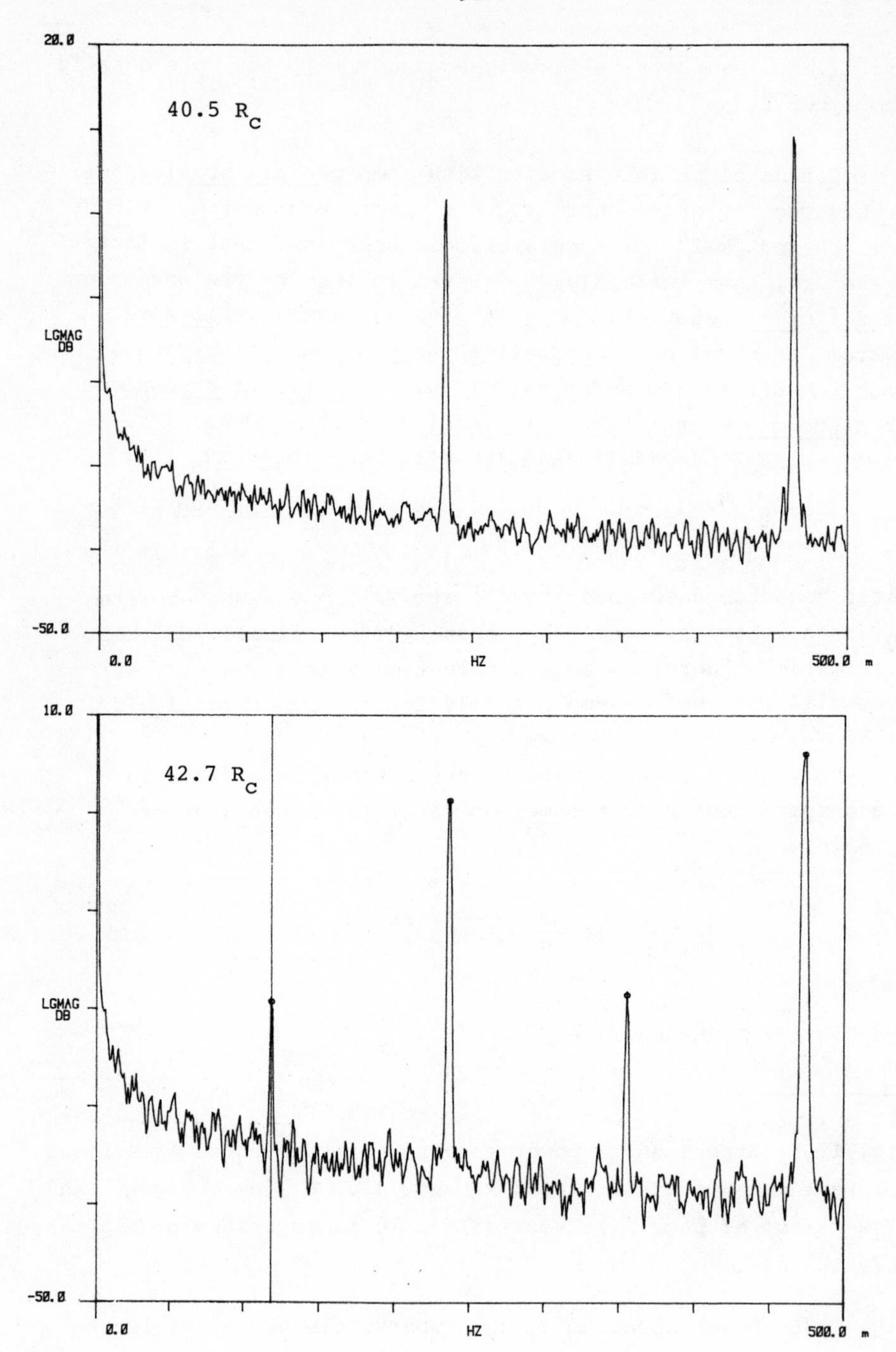

20.0
40.5 R_C
LGMAG DB
-50.0
0.0
HZ
500.0 m
10.0
42.7 R_C
LGMAG DB
-50.0
0.0
HZ
500.0 m

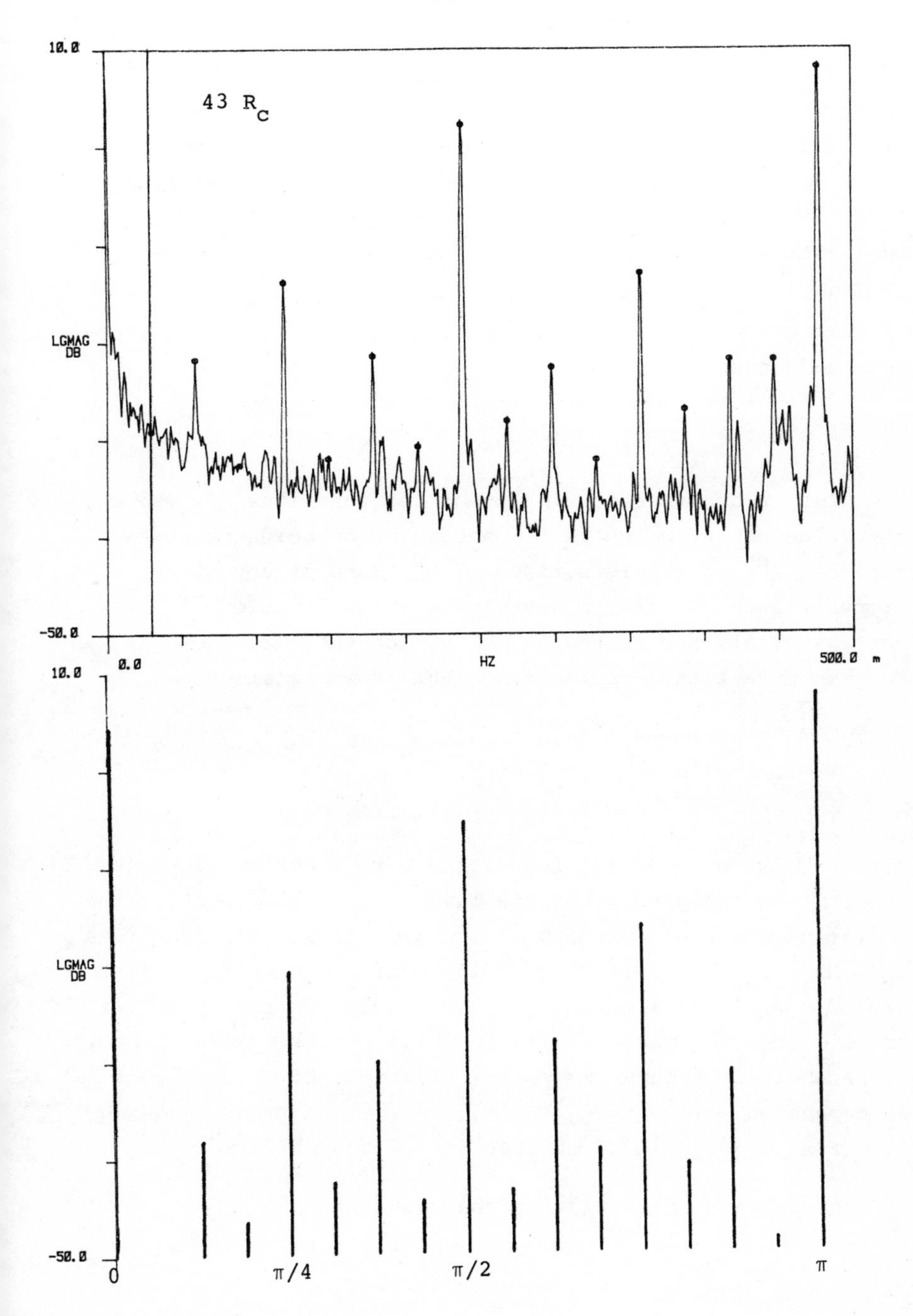
43 R_C
10.0
LGMAG DB
-50.0
0.0
HZ
500.0 m
10.0
LGMAG DB
-50.0
0
π/4
π/2
π

We have plotted a_k for the function f, to be described in Equation (*) below, and we are joining for comparison the measured graphs from the Libchaber-Maurer [1979] experiment. While the similarity is obvious, in our opinion a rigorous analytic treatment of the question is lacking. But some interesting partial results have been obtained by Feigenbaum [1979(1)],[1980]. See also the end of Section I.8 for a general discussion of the connection with hydrodynamical systems and equations.

Let us concentrate finally on the functions f_{μ_∞}, i.e., the value of μ which is the accumulation point of stable periods 2^n, of stable periods $3 \cdot 2^n$ (and of any other property such as ergodic behavior). This function f_{μ_∞} is not universal, nor is the value μ_∞ of the parameter, as can be seen by a little reflection. But if we consider

$$\lim_{m \to \infty} \frac{1}{\lambda^m} f_{\mu_\infty}^{2^m}(\lambda^m x) = f(x)$$

then f <u>is</u> a universal function--up to a change of scale--i.e., it is independent of the family f_μ. (λ is still the universal constant $-0.3995\ldots$.) Now f and all functions f_{μ_∞} have the histogram Fig. I.17, corresponding to a "stable periodic orbit of length 2^∞." The motion on the attractor is ergodic, but not mixing. The function f is an even function, it is analytic around $x = $ zero. If scaled to $f(0) = 1$, it has an expansion

$$f(x) \sim 1 - 1.52763x^2 + 0.104815x^4 - 0.0267057\ldots x^6 + \ldots .$$

By construction f satisfies the equation

$$f \circ f(\lambda x) = \lambda f(x) \tag{*}$$

and from $f(0) = 1$, we deduce that $\lambda = f(1)$. We find thus an <u>intrinsic characterization</u> of λ through a <u>nonlinear fixed point problem</u>. In the mathematical part, it is this fixed point equation which we take as a starting point for an analysis (in the spirit of the renormalization group) of one-parameter families of maps on the interval. The universal number δ is then also characterized intrinsically as the largest eigenvalue of the linear operator (on function space)

$$h(.) \rightarrow \frac{1}{\lambda} h(f(\lambda .)) + \frac{1}{\lambda} f'(f(\lambda .)) h(\lambda .) \quad .$$

I.7 HIGHER DIMENSIONAL SYSTEMS

As we have seen in Section I.1, there are many situations in which a map $\mathbb{R}^n \to \mathbb{R}^n$ seems more appropriate than a map $\mathbb{R}^1 \to \mathbb{R}^1$. The purpose of this section is to illustrate how the features of one-dimensional maps described in I.6 generalize to higher dimensions. We have already seen in the previous section that the numerical results seem to indicate a persistence of the universal behavior for higher dimensional systems. While we refer for the theorems to Section III.4, we want to illustrate here the salient features of their hypotheses in two dimensions. An often studied discrete dynamical system in two dimensions is the <u>Hénon map.</u> We write it here in the form

$$H_{b,\mu}\begin{pmatrix} x \\ y \end{pmatrix} = \begin{pmatrix} 1-\mu x^2+y \\ bx \end{pmatrix} ,$$

where μ and b are parameters. The Jacobian of this map is the matrix

$$\begin{pmatrix} -2\mu x & 1 \\ b & 0 \end{pmatrix} ,$$

and its determinant is $-b$. Therefore it is, for $0 < b < 1$ a map which contracts volumes and reverses orientation. Very loosely speaking, the contraction of the volumes will have the effect that the attractor (i.e., the set of points to which most points evolve under iterates of $H_{b,\mu}$) has at most one dimension. It seems to be thus general folklore to believe that $H_{b,\mu}$ should behave similarly to a one-dimensional map. Although we do not agree with this way of reasoning, we do agree with some of its conclusions.

To describe the result in a somewhat more precise fashion we recall the universal function f,

$$f(x) \sim 1 - 1.52763\ldots x^2 + \ldots ,$$

from Section I.6. As we have stated there, f is an analytic function of x^2. Therefore we can define

$$F\begin{pmatrix} x \\ y \end{pmatrix} = \begin{pmatrix} f\left(\sqrt{x^2-y}\right) \\ 0 \end{pmatrix}$$

as a map from $\mathbb{R}^2$ to $\mathbb{R}^2$. The result is now as follows: Collet-Eckmann-Koch [1980].

<u>Every one-parameter family of maps</u> $\mathbb{R}^2 \to \mathbb{R}^2$, <u>which passes sufficiently near to</u> F <u>will exhibit an infinite sequence of period-doubling bifurcations at values</u> μ_k. <u>The</u> μ_k <u>will have a limit</u> μ_∞ <u>and</u> $|\mu_n - \mu_\infty| \sim \text{const.}\ \delta^{-n}$.

It might seem an undue restriction to look only at F. But in fact the above result immediately generalizes to any family of maps which passes near F_T, where F_T is obtained through a coordinate transformation T from F, e.g.,

$$F_T\begin{pmatrix} x \\ y \end{pmatrix} = \begin{pmatrix} 2/3\ f\left(\sqrt{(2x+y)^2-x-2y}\right) \\ -1/3\ f\left(\sqrt{(2x+y)^2-x-2y}\right) \end{pmatrix}$$

when

$$T\begin{pmatrix} x \\ y \end{pmatrix} = \begin{pmatrix} 2x+y \\ x+2y \end{pmatrix} .$$

In particular the Hénon map, for small b passes near F, and this explains why one observes the universality (I.29, I.30) in this case. We insist that this is a nontrivial example, since the Hénon map is invertible, while the maps on the interval we have considered are not invertible. Note also that the scaling, which was λ in the one-dimensional case, is now given by <u>matrix</u> Λ, which for the example of F takes the form

$$\Lambda = \begin{pmatrix} \lambda & 0 \\ 0 & \lambda^2 \end{pmatrix}$$

while in the case of F_T it takes the form

$$\Lambda_T = \frac{1}{3} \begin{pmatrix} 4\lambda - \lambda^2 & 2\lambda - 2\lambda^2 \\ -2\lambda + 2\lambda^2 & -\lambda + 4\lambda^2 \end{pmatrix} .$$

We show this in the following figures. Define μ_M (the analog of the value μ_{M_0} for one-dimensional maps) to be that value of the parameter for which the unstable manifold (= invariant set tangent to unstable direction) at the fixed point P has a point of tangency with stable manifold at P (=invariant set tangent to stable direction). The picture is as in Figure I.37. (The stable and unstable directions at the fixed point P of the Henon map are defined as the eigendirections of the matrix $\begin{pmatrix} -2x\mu & 1 \\ b & 0 \end{pmatrix}$ at the value of x corresponding to P. Stable refers to the eigenvalue which is <1 in modulus, unstable to >1 in modulus). We then expect to find for $\mu \sim (\mu_M - \mu_\infty)/\delta + \mu_\infty$ ($\mu_\infty \sim 1.058...$ for Hénon when $b = 0.3$) the <u>same</u> image, but for the stable and unstable manifolds at the periodic points P_1 and P_2 of the period 2. See Figure I.38.

In addition, a rescaling of $-\lambda$ in the lateral and of $+\lambda^2$ in the vertical direction should give about the same picture as Fig. I.37. See Figure I.39. These relations get more pronounced for higher periods.

It is not known whether these points correspond to ergodic behavior on the attractor. It is, however known (Newhouse [1980]) that near these values of the parameter the Hénon map must have an <u>infinity</u> of <u>distinct</u> basins of attraction, i.e., an infinity of different initial conditions will stably converge to an infinity of different periodic orbits. This is

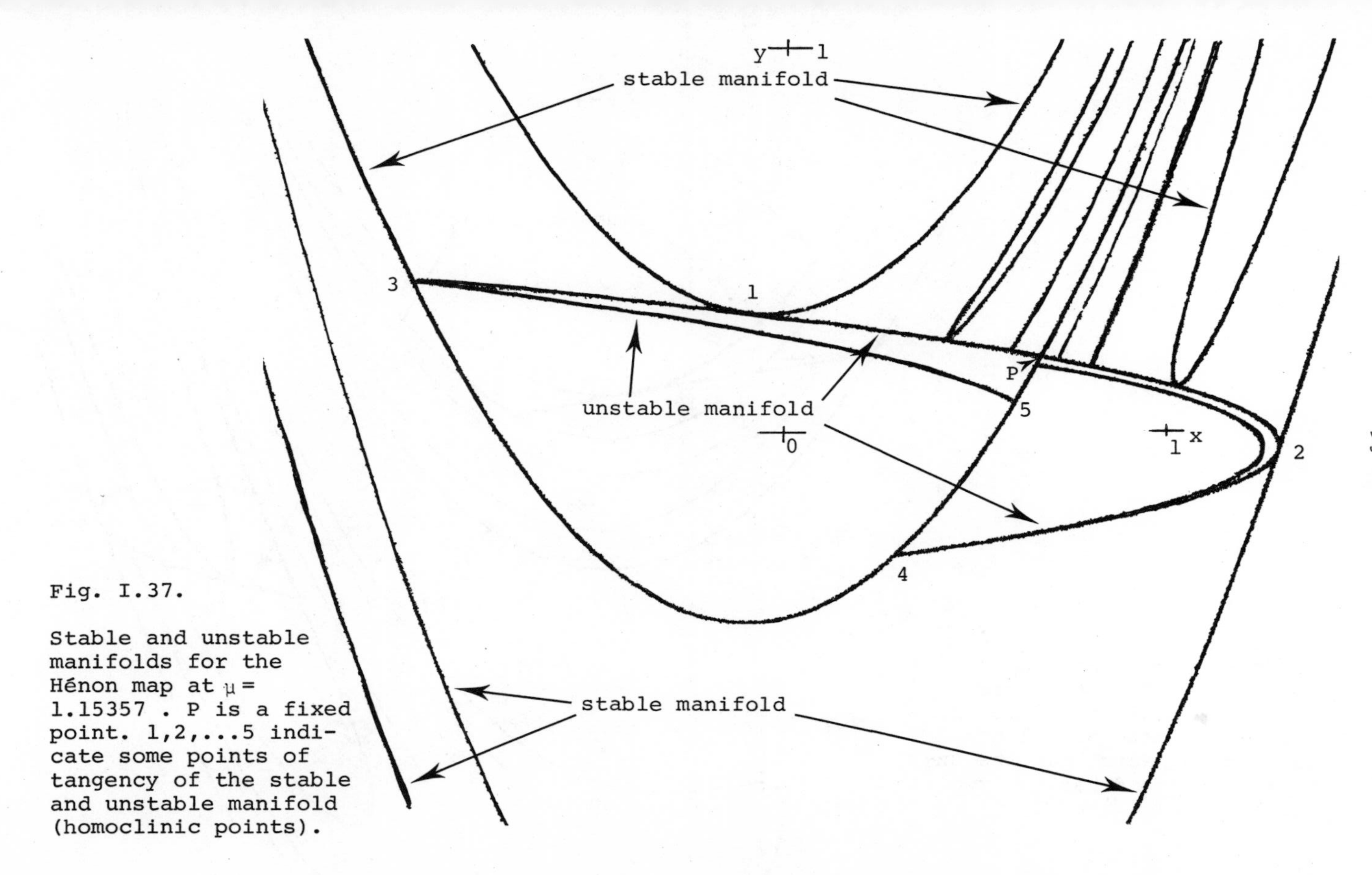

Fig. I.37.

Stable and unstable manifolds for the Hénon map at $\mu = 1.15357$. P is a fixed point. 1,2,...5 indicate some points of tangency of the stable and unstable manifold (homoclinic points).

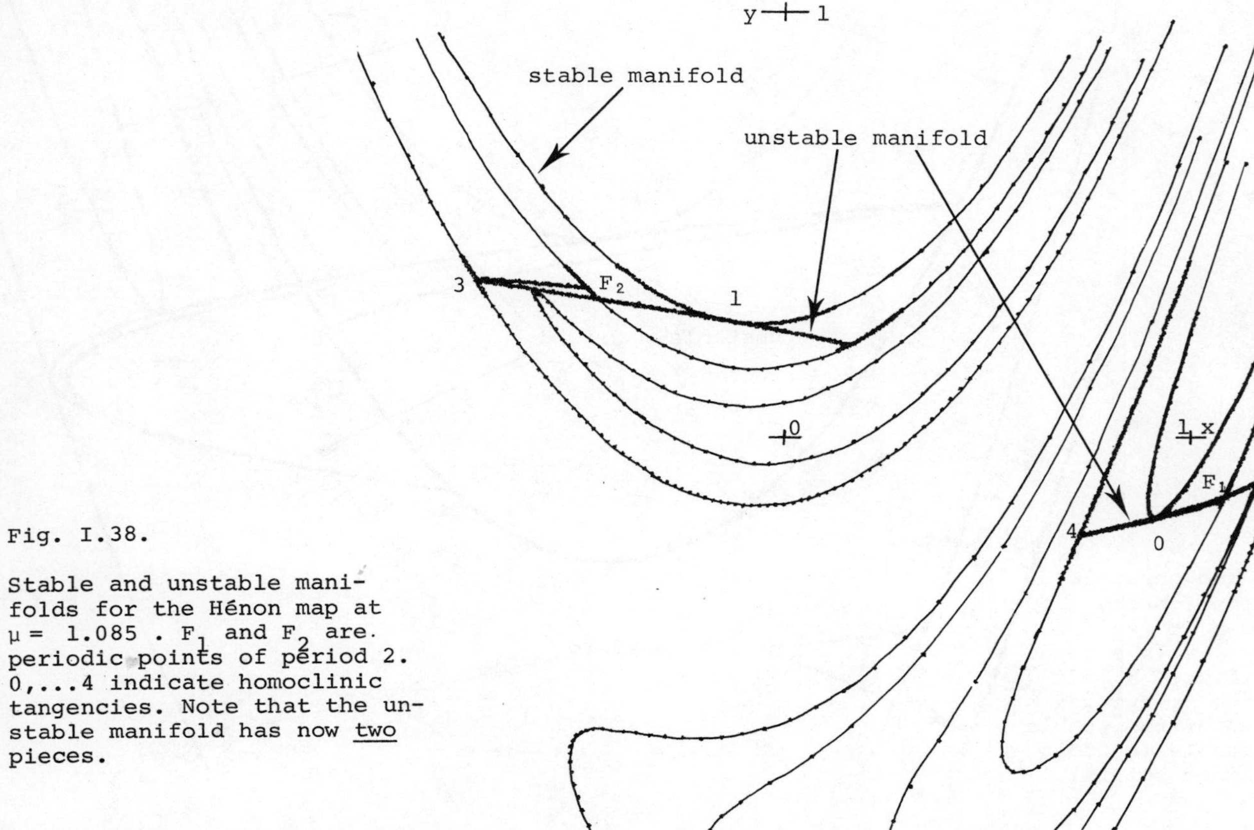

Fig. I.38.

Stable and unstable manifolds for the Hénon map at $\mu = 1.085$. F_1 and F_2 are periodic points of period 2. 0,...4 indicate homoclinic tangencies. Note that the unstable manifold has now _two_ pieces.

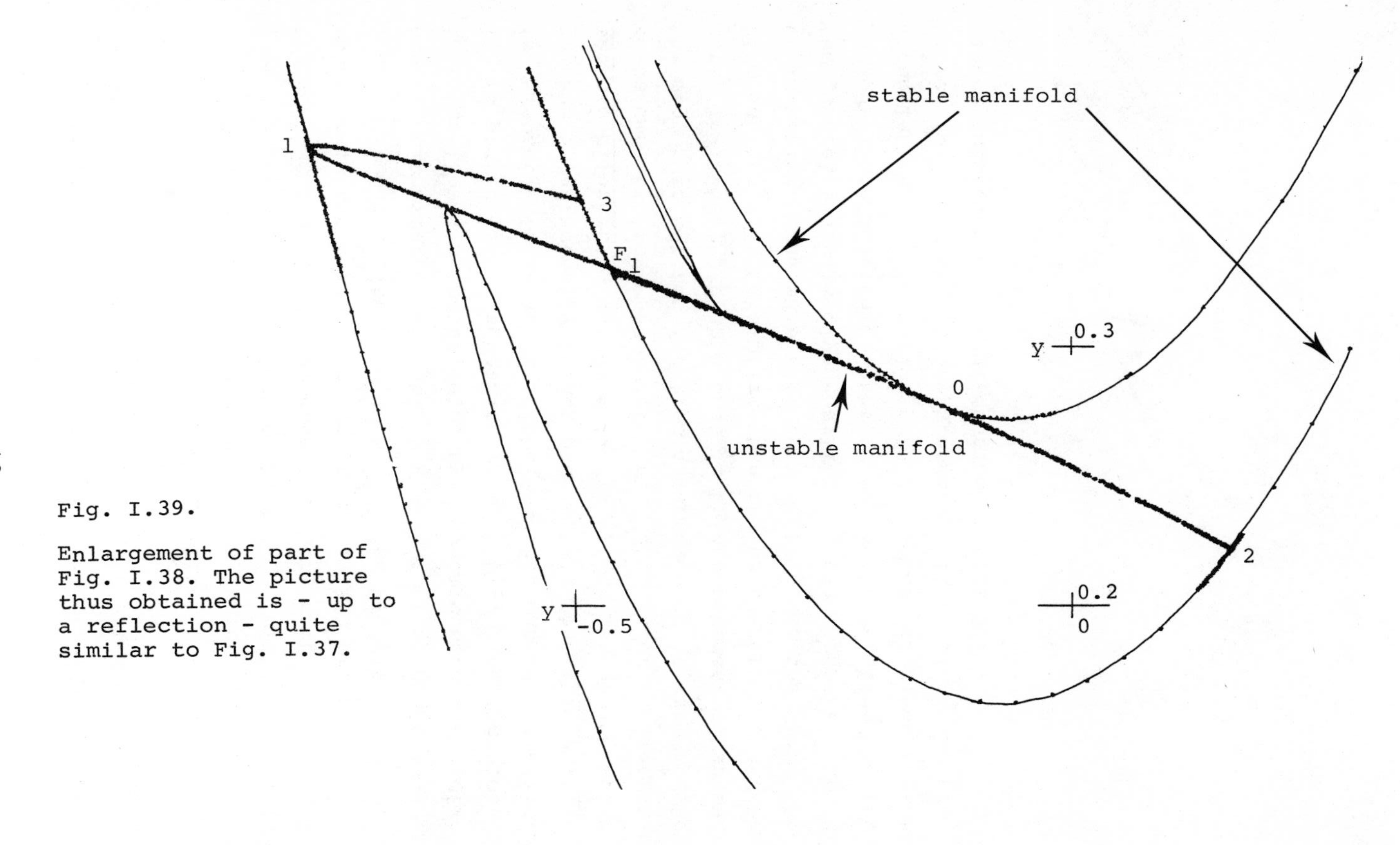

Fig. I.39.

Enlargement of part of Fig. I.38. The picture thus obtained is - up to a reflection - quite similar to Fig. I.37.

in marked contrast with the situation encountered for maps of the interval with negative Schwarzian derivative.

Let us analyze finally the Maurer-Libchaber experiment in the light of this extended universality (which holds, <u>with the same</u> δ for maps on $\mathbb{R}^n$ or maybe even on Hilbert manifolds). Assume that a differential equation such as the Navier-Stokes equation is the basis of hydrodynamics. The observation of a basic frequency (at $R = 40.5\,R_c$, see Fig. I.35) and of a subharmonic bifurcation (at $R = 42.7\,R_c$) strongly suggests that the flow (in the space of hydrodynamic modes, not of liquid particles) is something like in Fig. I.8. Therefore it is reasonable to assume that the Poincaré map P, mentioned on page 3 can actually be defined in this case. <u>Since we do see one bifurcation</u> (at $R = 42.7\,R_c$), it is very reasonable to assume that for the experimental setup under consideration, the map P is not too far from a map such as F above. Therefore, it is very probable that in this experiment subsequent bifurcations will take place when the parameter R, and hence P is varied. The same will then be true for the flow. By the general analysis given earlier one can then also expect periods 3 and aperiodic behavior on the "other side" of the accumulation of bifurcation points. In addition, approximate values for R where these phenomena will occur can be predicted according to our discussion on page 51.

PART II. PROPERTIES OF INDIVIDUAL MAPS

II.1 UNIMODAL MAPS AND THEIR ITINERARIES

We shall say that a mapping f of the interval $[-1,1]$ into itself is unimodal if

U1) f is continuous,

U2) $f(0) = 1$,

U3) f is strictly decreasing on $[0,1]$ and strictly increasing on $[-1,0]$.

We will say that f is $\mathscr{C}^1$-unimodal, if in addition

U4) f is once continuously differentiable, and $f'(x) \neq 0$ if $x \neq 0$.

Most of our considerations will be done for $\mathscr{C}^1$-unimodal functions, but in fact some results hold for unimodal maps, and some hold for any continuous map of an interval to itself. But the ideas of the proofs seem more transparent in the $\mathscr{C}^1$ case. It will also be easy to see that the particular choice of the interval $[-1,1]$ and fixing the extremum of f to be a maximum at $x = 0$ involves no loss of generality.

The topic of this book is the study of iterates of f: We shall denote $f^0 =$ identity, $f^1 = f$, $f^2 = f \circ f$, $f^n = f \circ f^{n-1}$, $n > 2$. The study of the iterates f^n is interesting if we consider the orbits $\mathcal{O}_f(x) = \{x, f(x), f^2(x), f^3(x), \ldots\}$ of points $x \in [-1,1]$.

A point $x \in [-1,1]$ is called periodic for f if $\mathcal{O}_f(x)$ is a finite set. The cardinality of this set is called the period of x, and $\mathcal{O}_f(x)$ is called the periodic orbit, of x. It is of course also the periodic orbit of $f^n(x)$, for all $n \geq 0$.

Our first aim is to study and classify orbits. A very interesting method has been advocated by Milnor and Thurston [1977]. It will lead later even to a topological classification of unimodal $\mathscr{C}^1$-maps, under mild additional assumptions. The results in the following sections II.2,3 are copies, variations or adaptions of known results. We have just combined ideas from the references Milnor-Thurston [1977], Metropolis-Stein-Stein [1973], Stefan [1977], Lanford [1979], Guckenheimer [1977], Derrida-Gervois-Pomeau [1979]. We hope this gives a coherent view.

Let f be a fixed unimodal mapping. We associate with $x \in [-1,1]$ a finite or infinite sequence of the symbols L,C,R (for left, center and right, respectively), called its <u>itinerary</u> $\underline{I}(x)$, as follows:

I1. $\underline{I}(x)$ is either an infinite sequence of L's and R's, or a finite (or empty) sequence of L's and R's, followed by C. The j-th element of $\underline{I}(x)$ will be denoted $I_j(x)$, $j = 0,1,\ldots$.

I2. If $f(x) \neq 0$, (0 is the maximum of f), for all $j \geq 0$, then $I_j(x) = L$ if $f^j(x) < 0$ and $I_j(x) = R$ if $f^j(x) > 0$.

I3. If $f^k(x) = 0$, for some k, then letting j denote the smallest such k, we set $I_j(x) = C$ and $I_\ell(x) = L$ if $0 \leq \ell < j$ and $f^\ell(x) < 0$ and $I_\ell(x) = R$ if $0 \leq \ell < j$ and $f^\ell(x) > 0$.

A sequence $\underline{I}$ of symbols L,C,R is called <u>admissible</u> if either $\underline{I}$ is an infinite sequence of L's and R's or if $\underline{I}$ is a finite (or empty) sequence of L's and R's, followed by C. Every itinerary is admissible. When dealing with sequences $\underline{I},\underline{J}$ we shall write $\underline{I}\underline{J}$ for the concatenation of $\underline{I}$ and $\underline{J}$, and $\underline{I}^n = \underline{I}\ldots\underline{I}$ (n times) and $\underline{I}^\infty = \underline{I}\underline{I}\ldots$ indefinitely. $|\underline{I}|$ will denote the cardinality of $\underline{I}$.

The <u>extended itinerary</u> $\underline{I}_E(x)$ is obtained by omitting the "stopping rules" in I1-I3. The definition of $\underline{I}_E(x)$ is

$\underline{I}_E(x) = \underline{I}(x)$ if $\underline{I}(x)$ is infinite,

$\underline{I}_E(x) = \underline{I}(x)\underline{I}(1)$ if $\underline{I}(x)$ is finite and $\underline{I}(1)$ is infinite,

$\underline{I}_E(x) = \underline{I}(x)(\underline{I}(1))^\infty$ if both $\underline{I}(x)$ and $\underline{I}(1)$ are finite.

Milnor-Thurston use "itinerary" to denote what we call "extended itinerary." The word <u>kneading sequence</u> is used for $\underline{I}(1)$. When we want to specify the function involved, we write $\underline{I}_f(1)$. The <u>shift operation</u> $\mathscr{S}$ will be used very often. If $\underline{I} = I_0 I_1 \ldots$ we define $\mathscr{S}\underline{I} = I_1 I_2 I_3 \ldots$. If $\underline{I} = C$, $\mathscr{S}$ is undefined, but we shall not mention this possibility every time it occurs in a proof. We write $\mathscr{S}^k$ for the k-fold iterate of $\mathscr{S}$. Note that $\mathscr{S}(\underline{I}(x)) = \underline{I}(f(x))$ unless $x=0$, and that $\mathscr{S}(\underline{I}_E(x)) = \underline{I}_E(f(x))$ (always).

We shall now introduce an <u>ordering</u> $<$ <u>between different admissible sequences</u>, and this ordering will be such that $x < y \Rightarrow \underline{I}(x) \leq \underline{I}(y)$ and $\underline{I}(x) < \underline{I}(y) \Rightarrow x < y$. Thus the order of the real line will be reflected in the order of the itineraries. Here the fact that we are dealing with one-dimensional dynamical systems is crucial. We proceed to define the ordering. First of all, we shall say $L < C < R$. Let now $\underline{A} \neq \underline{B}$ be two admissible sequences. Let i be the first index for which $A_i \neq B_i$. Note that if $\ell = |\underline{A}| < |\underline{B}|$, $\ell < \infty$, then, since $\underline{A},\underline{B}$ are admissible, such an index will exist since $A_{\ell-1} = C$ and $B_{\ell-1} \neq C$. In the other cases the existence of such an index is clear. We say that $\underline{A} < \underline{B}$ if either

$\mathscr{O}$1. There are an even number of R's in $A_0 A_1 \ldots A_{i-1} = B_0 B_1 \ldots B_{i-1}$ and $A_i < B_i$

or

$\mathcal{O}2$. There are an odd number of R's in
$A_0A_1\ldots A_{i-1}$ and $A_i > B_i$.

If none of these hold then we say $\underline{B} < \underline{A}$.

We shall use the notation $\underline{A} \le \underline{B}$, $\underline{A} \ge \underline{B}$, $\underline{A} > \underline{B}$ in the standard way. It will be useful to define a finite sequence $\underline{B}$ to be _even_ if it has an even number of R's, _odd_ otherwise. Even and odd will never refer to the number of elements $n = |\underline{B}|$ unless specifically stated.

LEMMA II.1.1. _The relation_ $<$ _is a complete ordering._

Proof. This is trivial to check.

We have now the easy, but important relations:

LEMMA II.1.2. _If_ f _is unimodal and_ $\underline{I}(x) < \underline{I}(x')$ _then_ $x < x'$.

LEMMA II.1.3. _If_ f _is unimodal and_ $x < x'$ _then_ $\underline{I}(x) \le \underline{I}(x')$.

Proof of Lemma II.1.2. By induction on the smallest index i for which $I_i(x) \neq I_i(x')$. The statement is true when $i = 0$. Suppose $i > 0$, and suppose we have already shown the result for $i - 1$. When $I_0(x) = I_0(x') = L$, then if $\underline{I}(x) < \underline{I}(x')$ we find $\underline{I}(f(x)) < \underline{I}(f(x'))$. But $\underline{I}(f(x)) = I_1(x)I_2(x)\ldots$, and hence $I_{i-1}(f(x)) \neq I_{i-1}(f(x'))$. By the induction hypothesis $f(x) < f(x')$. Since $I_0(x) = I_0(x') = L$, we have $x, x' \in [-1,0)$ and since $f|_{[-1,0]}$ is strictly increasing, $f(x) < f(x')$ implies $x < x'$.

When $I_0(x) = I_0(x') = R$, then if $\underline{I}(x) < \underline{I}(x')$ we find $\underline{I}(f(x)) > \underline{I}(f(x'))$ since $\underline{I}(f(x))$ has one R less than $\underline{I}(x)$ before i. Hence $f(x) > f(x')$. Since $I_0(x) = I_0(x') = R$, we have $x, x' \in (0,1]$ and since $f|_{[0,1]}$ is strictly decreasing, $f(x) > f(x')$ implies $x < x'$.

Proof of Lemma II.1.3. This is obvious from the fact that $<$ is a complete ordering. Namely if $x < x'$ and $\underline{I}(x) > \underline{I}(x')$ then Lemma 2 implies a contradiction.

REMARK II.1.4. We stress here a quite general observation about iterates of unimodal maps. Consider the set of points $\{x \mid f^k(x) = 0 \text{ for some } k, \ 0 \le k < n\}$. Since f is unimodal, this set is finite. These points divide the interval $[-1,1]$ into a finite number of subintervals. We claim f^n is strictly monotone on each of these subintervals. Cf. Fig. II.1.

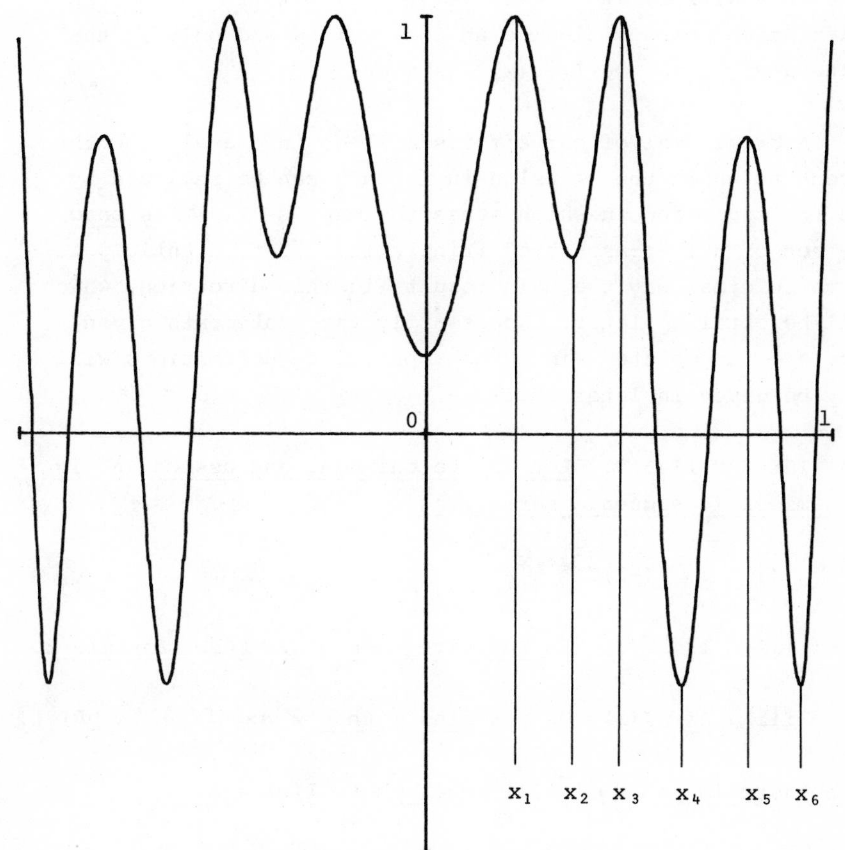

Figure II.1. $f(x) = 1 - 1.6x^2$, shown is $f^5(x)$, $f^4(x_1) = f^4(x_3) = f^3(x_4) = f^3(x_6) = f^2(x_2) = f^1(x_5) = f^0(0) = 0$
Example: For $y \in (x_1, x_2)$, $\underline{I}(y) = RRLRL\ldots$

<u>Proof</u>. Let x, x' be two points strictly inside one of the intervals. Their itineraries must coincide for the indices $0,1,2,\ldots,n-1$. If not, there is some smallest $j < n$ such that, say $f^j(x) < 0$ and $f^j(x') > 0$. Then by continuity there must be a y between x and x' for which $f^j(y) = 0$, a contradiction. But if the itineraries coincide, f^n is a homeomorphism, since f is a homeomorphism on any interval which does not contain zero in its interior.

In addition, it is easy to verify that f^n preserves orientation near x inside an interval if and only if the itinerary $I_0(x) \ldots I_{n-1}(x)$ is even.

A great deal of our efforts will go into analyzing the extent to which the $\leq$ sign in Lemma 3 can be replaced by the $<$ sign, and in which sense the map $x \to \underline{I}(x)$ is <u>onto</u> the set of all "reasonable" itineraries. Let us anticipate in an informal way the main results in this direction, which will be obtained later. Necessarily the statements given here are incomplete, since the appropriate definitions will only be given in later sections.

THEOREM II.3.8. <u>Let</u> f <u>be unimodal and assume</u> $\underline{A}$ <u>is an admissible sequence satisfying</u>

(i) $\underline{A} \geq \underline{I}(-1)$.

(ii) <u>If</u> $\underline{I}(1)$ <u>is infinite then</u> $\mathscr{S}^k\underline{A} < \underline{I}(1)$ <u>for all</u> k.

(iii) <u>If</u> $\underline{I}(1) = \underline{D}C$ <u>is finite then</u> $\mathscr{S}^k\underline{A} < \inf((\underline{D}L)^\infty, (\underline{D}R)^\infty)$.

<u>Then there is an</u> $x \in [-1,1]$ <u>such that</u> $\underline{I}(x) = \underline{A}$.

Thus the map $x \to \underline{I}(x)$ is onto in this sense. The question of whether $x \to \underline{I}(x)$ is 1-1 will turn out to be much harder to analyze. Nevertheless, the result is very easy to formulate.

THEOREM II.5.4. *Let f be S-unimodal and assume f has no stable periodic orbit. Then* $\underline{I}(x) < \underline{I}(y) \Longleftrightarrow x < y$.

The term S-unimodal denotes $\mathscr{C}^1$-unimodal functions f with negative Schwarzian derivative, mapping $[f(1),1]$ into itself. Every function of the form $f(x) = 1 - \mu x^2$, with $0 < \mu \leq 2$ is S-unimodal. Once the theory will be developed to this point, it will be relatively easy, although sometimes a little lengthy to see that for the class of S-unimodal functions f, the quantity $\underline{K}(f) = \underline{I}_f(1)$ (the kneading sequence) is in fact almost a topological invariant. We pick three important results which are easy to formulate at this stage.

THEOREM. *If f is S-unimodal then*

(i) f *has a stable periodic orbit if and only if* $\underline{K}(f)$ *is periodic or finite (not only eventually periodic).*

(ii) *If* $\underline{K}(f)$ *is not periodic and* $\underline{K}(g) = \underline{K}(f)$ *then* $f|_{[f(1),1]}$ *and* $g|_{[g(1),1]}$ *are topologically conjugate.*

(iii) *If* $\underline{K}(f)$ *is eventually periodic, but not periodic, then* $f|_{[f(1),1]}$ *has an invariant measure which is absolutely continuous with respect to Lebesgue measure.*

These results show that the notion of negative Schwarzian derivative was instrumental to go from the weaker results of *Milnor-Thurston* (who pioneered the field of itineraries) to the specific results of *Guckenheimer* ((i), (ii) above) and of *Misiurewicz* (iii). The results can be found as follows (i) = I.6.2, (ii) = I.6.1, (iii) = I.8.1. The role of the negative Schwarzian derivative was discovered by *Singer*, who showed that an S-unimodal map can have at most one stable periodic orbit (a conclusion which is wrong for general $\mathscr{C}^1$-unimodal maps).

<u>Remarks and Bibliography</u>. Iterations of maps have been extensively studied by Julia [1918] and Fatou [1919]. Most often, these maps were either defined by analytic functions, or one considered the map $x \to 1/x - [1/x]$ occurring in the study of continuous fractions.

The coding of itineraries has a long past, whose origin we were not able to trace, cf. Morse [1966]. For unimodal maps, the idea is clearly present in Sarkovskii [1964], Stefan [1977], Metropolis-Stein-Stein [1973]. The reference of Milnor-Thurston [1977] goes much further than the references above, in two respects. Firstly, it generalizes the ideas to maps with several intervals of monotony. Secondly, it contains a wealth of ideas on topological conjugacy and zeta-functions. The papers of Guckenheimer [1979] and Misiurewicz [1980], which we shall discuss in great detail in II.5-8, are elaborations of these ideas.

II.2 THE CALCULUS OF ITINERARIES

We shall now leave maps for a moment, and draw some combinatorial conclusions from the definition of the ordering of itineraries. The definitions given below are of course motivated through the connections between itineraries and maps. We shall indicate these connections for illustration, but they do not enter our arguments.

Fix a unimodal f and consider a periodic orbit $\{x, f(x), \ldots, f^{n-1}(x)\}$ of period n. Each of the points $f^j(x)$ has a periodic extended itinerary $\underline{I}_E(f^j(x))$. In some sense they are all equivalent, since they represent the same periodic orbit. We want to single out the maximal $\underline{I}_E(f^j(x))$. It is, by Lemma 1.3, $\underline{I}_E(y)$, where $y = \max_{0 \le j < n} f^j(x)$. Then all shifts $\mathscr{S}^k(\underline{I}_E(y))$ satisfy $\mathscr{S}^k(\underline{I}_E(y)) \le \underline{I}_E(y)$. This motivates the following definitions.

We call a sequence $\underline{I}$ <u>maximal</u> if it is admissible and if $\mathscr{S}^k(\underline{I}) \le \underline{I}$ for $k = 1, 2, \ldots, |\underline{I}| - 1$, when $\underline{I}$ is finite and for $k = 1, 2, \ldots$ when $\underline{I}$ is infinite.

Note that if f is unimodal, $\underline{I}(1)$ is always maximal. If $\underline{I}$ is <u>finite and maximal</u> then it follows that

$$\mathscr{S}^k(\underline{I}) < \underline{I} \quad \text{for} \quad k = 1, 2, \ldots, |\underline{I}| - 1 \ .$$

LEMMA II.2.1. <u>The only maximal sequences not starting with</u> RL... <u>are</u> L^∞, R^∞ RC <u>and</u> C.

<u>Proof</u>. Left to the reader.

The following situation, see below for an explicit example, is of special interest to us. Assume f is unimodal and assume $\underline{I}_f(1) = \underline{A}L\underline{A}\ldots$, if $\underline{A}$ is odd, or $\underline{I}(1) = \underline{A}R\underline{A}\ldots$

if $\underline{A}$ is even. Let $n = |\underline{A}| + 1$ and define $a = f^n(0)$. Let now $J_0 = [-|a|,|a|]$. Our main assumptions on f are (H ∿ "harmonics")

H1. $f^n(J_0) \subset J_0$

H2. $f^k(J_0) \cap f^\ell(J_0) = \emptyset$ for all $k,\ell, 0 \le k < \ell < n$.

Then, since f is unimodal, $g(y) = 1/a\ f^n(ay)|_{J_0}$ is unimodal, too. The *-product which we are going to define now is devised to describe the relations between the itineraries for f and g. Let $\underline{A}$ be a finite nonempty sequence of L's and R's and let $\underline{B}$ be admissible. We define $\underline{A}*\underline{B}$ as follows:

P1. If $\underline{A}$ is even and $\underline{B}$ is infinite, then
$\underline{A}*\underline{B} = \underline{A}B_0\underline{A}B_1\underline{A}B_2\ldots$.

P2. If $\underline{A}$ is even and $\underline{B} = B_0 \ldots B_{n-1}C$ is finite, then
$\underline{A}*\underline{B} = \underline{A}B_0\underline{A}B_1\underline{A}\ldots\underline{A}B_{n-1}\underline{A}C$.

P3. If $\underline{A}$ is odd and $\underline{B}$ is infinite, then
$\underline{A}*\underline{B} = \underline{A}\check{B}_0\underline{A}\check{B}_1\underline{A}\check{B}_2\ldots$, where $\check{L} = R$, $\check{R} = L$, $\check{C} = C$.

P4. If $\underline{A}$ is odd and $\underline{B} = B_0 \ldots B_{n-1}C$ is finite, then
$\underline{A}*\underline{B} = \underline{A}\check{B}_0\underline{A}\check{B}_1\underline{A}\ldots\underline{A}\check{B}_{n-1}\underline{A}C$.

We illustrate the meaning of the <u>*-product</u> for an example with f,g of the type described above.

EXAMPLE. $f(x) = 1 - 1.476\ldots x^2$.

For this function, $\underline{I}_f(1) = RLRRRC = R*RLC$. See the Figures II.2, II.3 below. $f \circ f$ maps J_0 into itself and $f \circ f$ restricted to $[f(1),-f(1)]$ is anti-unimodal, i.e., $g(y) = f(1)^{-1} f \circ f(f(1)y)|_{J_0}$ is unimodal.

Consider the itinerary $\underline{I}_g(y)$ for $y \in [-1,1]$. If $y > 0$ then $f(1)y < 0$ and if $g(y) > 0$ then $f \circ f(f(1)y) < 0$. Thus

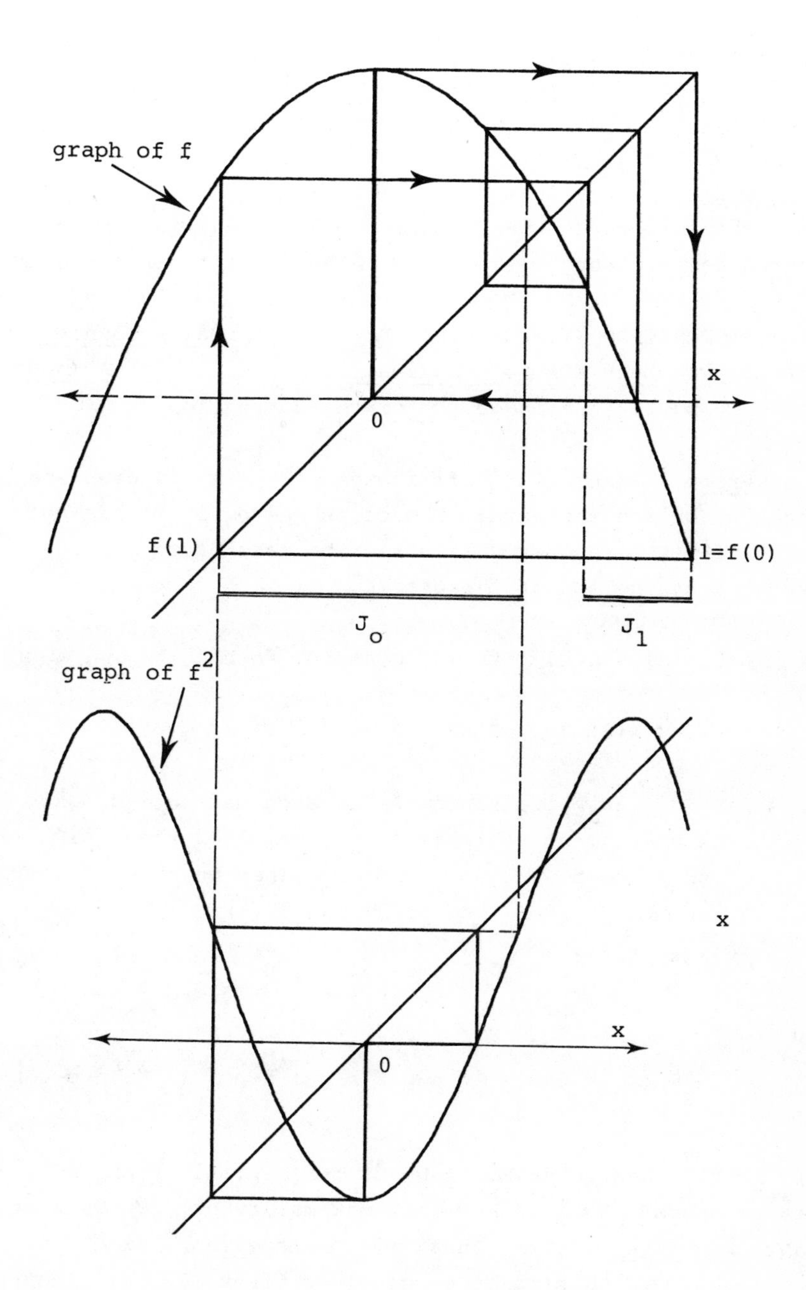

Fig. II.2, Fig. II.3. A map f and its square f∘f.

$\underline{I}_g(y) = \check{\underline{I}}_{f\circ f}(f(1)y)$. But this means that $\underline{I}_f(f(1)y) = \check{I}_{f\circ f,0}(f(1)y)R\check{I}_{f\circ f,1}(f(1)y)R\ldots$ since J_0 is mapped by f into J_1, and $x \in J_1$ implies $x > 0$. Thus $R\underline{I}_f(f(1)y) = R*\underline{I}_g(y)$. The general formula under the assumptions H1, H2 is $\underline{A}\underline{I}_f(f^n(0)y) = \underline{A}*\underline{I}_g(y)$, where $n = |\underline{A}|+1$.

The following two propositions show that the *-product introduces a strong systematics of the ordering of itineraries.

PROPOSITION II.2.2. _Let_ $\underline{B}$ _be a maximal sequence and let_ $\underline{B} < \underline{D}$. _Let_ $\underline{A}C$ _be a maximal sequence. Then_ $\mathscr{S}^k(\underline{A}*\underline{B}) < \underline{A}*\underline{D}$ _for all_ $k \geq 0$. (See page 65 for the definition of $\mathscr{S}^k$.)

Proof. Assume for definiteness that $\underline{A}$ is even and that $\underline{B}$, $\underline{D}$ are infinite. The cases where $\underline{B}$, $\underline{D}$ are not both infinite are easy variants. The case of odd $\underline{A}$ will be sketched at the end of the proof.

Let $|\underline{A}| = n-1$. If $k = nm$, $m \geq 0$, then

$$\mathscr{S}^k(\underline{A}*\underline{B}) = \underline{A}B_m\underline{A}B_{m+1}\underline{A}\ldots \quad .$$

This is $\leq \underline{A}B_0\underline{A}B_1\underline{A}\ldots$ since $\underline{A}$ is even and $\underline{B}$ is maximal. By the same argument, this is in turn smaller than $\underline{A}*\underline{D}$. So let now $k = mn+p$, $m \geq 0$, $1 \leq p < n$. Then we have to compare

$$\mathscr{S}^k(\underline{A}*\underline{B}) = A_pA_{p+1}\ldots A_{n-2} \quad B_m \quad A_0 \quad \ldots = \underline{X}\; B_m \quad A_0 \quad \ldots$$

to

$$\underline{A}*\underline{D} = A_0 \quad \ldots A_{n-p-2}A_{n-p-1}A_{n-p}\ldots = \underline{X}'A_{n-p-1}A_{n-p}\ldots$$

If $\underline{X} \neq \underline{X}'$, then $\mathscr{S}^k(\underline{A}*\underline{B}) < \underline{A}*\underline{D}$ by the maximality of $\underline{A}C$. So let us assume $\underline{X} = \underline{X}'$. From the maximality of $\underline{A}C$ we must have $\underline{X}C < \underline{X}A_{n-p-1}\ldots$. Thus, if $\underline{X}$ is <u>even</u> we find $A_{n-p-1} = R$. In this case, we consider first $B_m = R$. Then $\underline{X}B_m$ is odd, and the assertion will follow from

$$\underline{Y}B_{m+1}\ldots \equiv A_0\ldots A_{n-2}B_{m+1}\ldots > A_{n-p}\ldots A_{n-2}D_0\ldots$$

$$\equiv \underline{Y}'D_0\ldots \quad . \qquad (*)$$

Since $\underline{B}$ is maximal, and $\underline{D} > \underline{B}$ we have $D_0 \geq C$, by Lemma 1. Since $\underline{X}'R$ is odd and $\underline{A}$ is even, we find that $\underline{Y}'$ is odd. Thus

$$\underline{Y}'D_0\ldots \leq \underline{Y}'C = \mathscr{S}^{n-p}(\underline{A}C) < \underline{A}\ldots ,$$

since $n-p \geq 1$. Thus (*) follows in this case. Assuming still that $\underline{X}$ is even, we consider now $B_m = L$. But then $\underline{X}L\ldots < \underline{X}'A_{n-p-1}\ldots = \underline{X}'R\ldots$. The assertion of the proposition follows in this case.

The case of odd $\underline{X}$ is handled by analogy. Let us consider in part the proof for odd $\underline{A}$. We observe:

LEMMA II.2.3. *If* $\underline{A}$ *is odd,* $I_i \neq C$, $i \leq n$, *then* $\underline{A}\check{I}_0\underline{A}\check{I}_1\ldots\underline{A}\check{I}_n$ *is even iff* $I_0\ldots I_n$ *is even.*

The proof is left to the reader. HINT: Distinguish the cases of even and odd n. The proof of the proposition for the case of odd $\underline{A}$ is now an easy variant of what we have done for even $\underline{A}$.

REMARK. We have shown in fact that if $\underline{B} = \underline{D}$, then $\mathscr{S}^k(\underline{A}*\underline{B}) \leq \underline{A}*\underline{B}$ for $k \geq 0$. Thus we have the two important consequences.

COROLLARY II.2.4. *If* $\underline{A}C$ *and* $\underline{B}$ *are maximal, then* $\underline{A}*\underline{B}$ *is maximal.*

THEOREM II.2.5. *If* $\underline{A}C$ *is maximal then the map* $\underline{A}*:\underline{B} \mapsto \underline{A}*\underline{B}$ *is order preserving from the set of maximal itineraries to itself.*

Note that $\underline{A}*C = \underline{A}C$, $\emptyset*\underline{B} = \underline{B}$.

LEMMA II.2.6. *If* $\underline{A}_1 C$ *and* $\underline{A}_2 C$ *are finite and* $\underline{B}$ *is admissible, then one has*

$$\underline{A}_1*(\underline{A}_2*\underline{B}) = \underline{A}*\underline{B} \quad ,$$

where $\underline{A}C = \underline{A}_1*\underline{A}_2 C$.

The proof is easy and left to the reader.

In view of this lemma the notation $\underline{A}*^n$ for the n-fold *-product is justified.

The next result is very strong. It says that $\underline{A}*$ is in some sense *onto*.

THEOREM II.2.7. *Let* $\underline{A}C$ *and* $\underline{Q} \neq C$ *be maximal, and set* $n = |\underline{A}C|$. *Assume* $\underline{A}*\mathscr{S}\underline{Q} \leq \underline{X} \leq \underline{A}*\underline{Q}$,

(i) *If* $\underline{X}$ *is maximal then there is a unique* $\underline{B}$ *such that* $\underline{X} = \underline{A}*\underline{B}$. *This* $\underline{B}$ *is maximal*.

(ii) *If* $\underline{X}$ *is admissible and if for all* $p \geq 1$, $\mathscr{S}^{np}\underline{X} \leq \underline{A}*\underline{Q}$, *then there is a unique* $\underline{B}$ *such that* $\underline{X} = \underline{A}*\underline{B}$.

(iii) *In either case* (i) *or* (ii) *one has*

$$\underline{A}*\mathscr{S}\underline{Q} \leq \mathscr{S}^{np}\underline{X} \leq \underline{A}*\underline{Q}$$

for all $p \geq 1$.

Proof. If $\underline{Q} = R^\infty$ or L^∞ the assertions of the theorem are obvious. So assume now $\underline{Q} \neq R^\infty$ and $\neq L^\infty$. We first show, in the case $\underline{Q} \neq RC$ that

$$\underline{A}*\mathscr{S}\underline{Q} \leq \underline{A}*\mathscr{S}^2\underline{Q} \leq \underline{A}*\underline{Q} \qquad (*)$$

If $\underline{Q} = RL^{\infty}$ the assertion (*) is obvious. In all other cases $\underline{Q} = RL^{t}T\ldots$ with $T = R$ or C and no substring of $\underline{Q}$ can be of the form $\ldots RL^{s}T'\ldots$ with $s > t$. Assume for definiteness that $\underline{A}$ is even. By inspection, the assertion (*) follows from

$$\begin{aligned}\underline{A}*\mathscr{S}\underline{Q} &\le \underline{A}Q_1\underline{A}Q_2\ldots\underline{A}Q_t\underline{A}Q_{t+1}\ldots = (\underline{A}L)^{t}\underline{A}T\ldots\\ &< (\underline{A}L)^{t-1}\underline{A}T\ldots = \underline{A}Q_2\ldots = \underline{A}*\mathscr{S}^{2}\underline{Q}\\ &\le \underline{A}R(\underline{A}L)^{t}\ldots = \underline{A}*\underline{Q}\ .\end{aligned}$$

Proof of Theorem II.2.7(iii). By induction, it is sufficient to prove the case $p = 1$. Assume for definiteness that $\underline{A}$ is even. We clearly must have $\underline{X} = \underline{A}B_o\ldots$. Consider now $\mathscr{S}^{n}\underline{X}$, and assume first $\mathscr{S}^{n}\underline{X} \le \underline{X}$. In this case it suffices to prove $\mathscr{S}^{n}\underline{X} \le \underline{A}*\mathscr{S}\underline{Q}$. This assertion is obvious when $\underline{Q} = RC$ since $B_o = C$ or $B_o = R$. In the remaining cases, $\underline{Q} = RL\ldots$. If $B_o = C$ the assertion is again obvious. If $B_o = L$, we find, since $\underline{A}L$ is even, that $\underline{A}*\mathscr{S}\underline{Q} \le \underline{X}$ implies

$$\mathscr{S}^{n}\underline{X} \ge \underline{A}*\mathscr{S}^{2}\underline{Q}$$

and the assertion follows from (*) above. Finally if $B_o = R$, then $\underline{A}B_o$ is odd and hence $\underline{X} \le \underline{A}*\underline{Q}$ implies $\mathscr{S}^{n}\underline{X} \ge \underline{A}*\mathscr{S}\underline{Q}$. Consider next the case $\mathscr{S}^{n}\underline{X} > \underline{X}$. We have to show $\mathscr{S}^{n}\underline{X} \le \underline{A}*\underline{Q}$. If $\underline{X}$ is not maximal, this is part of the assumption and if $\underline{X}$ is maximal $\mathscr{S}^{n}\underline{X} > \underline{X}$ is impossible. So (iii) is proven.

Proof of (i), (ii). From

$$\underline{A}*\mathscr{S}\underline{Q} \le \underline{X} \le \underline{A}*\underline{Q}$$

it follows that $\underline{X}$ must be of the form $\underline{A}B_0\hat{\underline{X}}$, and from (iii) it follows inductively that $\hat{\underline{X}}$ must be again of the same form. Hence $\underline{X} = \underline{A}*\underline{B}$. The uniqueness is obvious and the maximality in case (i) is obvious.

We now analyze in more detail periodic sequences of L's and R's, and we ask for their "first" appearance with respect to the ordering $<$. To formulate this question precisely, consider all periodic sequences of L's and R's which have period $p \geq 2$. This is, for fixed p, a finite set of sequences (at most 2^p). Let $\underline{B}^{\infty}_{1,p} < \underline{B}^{\infty}_{2,p} < \ldots < \underline{B}^{\infty}_{n(p),p}$ be an enumeration of the maximal elements in this set, with $|\underline{B}_{i,p}| = p$. We call $\underline{B}_{1,p}$ the min-max of period p, and we shall reserve the symbol $\underline{P}_p$ for this.

The ordering of the $\underline{P}_p$ for different p will be the combinatorial basis of the Šarkovskii theorem in Section 3.

Define the following ordering of the integers ≥ 1.

$$\begin{aligned}
&3 > 5 > 7 > 9 > \ldots\\
&> 2\cdot 3 > 2\cdot 5 > 2\cdot 7 > 2\cdot 9 > \ldots\\
&\ldots\\
&> 2^n\cdot 3 > 2^n\cdot 5 > 2^n\cdot 7 > \ldots\\
&\ldots\\
&> \ldots > 2^m > \ldots > 16 > 8 > 4 > 2 > 1 \quad .
\end{aligned}$$

That is, first the odd integers ≥ 3, then the powers of 2 times the odd integers, and then the powers of 2 backwards.

THEOREM II.2.8. (Pre-Šarkovskii Theorem). *Let* s,t *be two integers. If* $s < t$ *in the sense of the above ordering, then* $\underline{P}_s < \underline{P}_t$.

This result will follow from the stronger result, where we identify the $\underline{P}_i$.

THEOREM II.2.9. *The min-max* $\underline{P}_i$ *are of the following form.*

1. *If* $i \geq 3$ *is odd,* $\underline{P}_i = RLR^{i-2}$.
2. *If* $i = 2^n\cdot k$, *where* $k \geq 3$ *is odd, then* $\underline{P}_i = R^{*n} * RLR^{k-2}$.

3. <u>If</u> $i = 2^n$, $n > 0$, <u>then</u> $\underline{P}_i = R^{*n}*R$, <u>and</u> $\underline{P}_1 = L$. (We have extended somewhat the notation *.)

REMARK. By Lemma 3, $\underline{P}_i$ is even for all i of the form $2^n k$, $k \geq 3$ odd, and odd for $i = 2^n$, $n \geq 1$.

The proofs of Theorems 8, 9 proceed through a series of lemmas which we list now.

LEMMA II.2.10. <u>If</u> $p \geq 3$ <u>is odd, then</u> $\underline{P}_p = RLR^{p-2}$.

LEMMA II.2.11. <u>For all</u> $n \geq 1$ <u>and all odd</u> $k \geq 2$,

$$R^{*n}*(RLR^{k-2})^{\infty} < R^{*(n-1)}*(RLR^{k-2})^{\infty} ,$$

<u>where</u> $R^{*0} = \emptyset$.

LEMMA II.2.12. <u>The sequences</u> $R^{\infty} < RC < R*R^{\infty} < R*RC < \ldots < R^{*n}*R^{\infty} < R^{*n}*RC < R^{*(n+1)}*R^{\infty} < \ldots$ <u>are consecutive among the maximal sequences</u>: <u>If</u> $\underline{X}$ <u>is maximal and</u> $R^{*n}*RC \leq \underline{X} < R^{*(n+1)}*R^{\infty}$, <u>then</u> $\underline{X} = R^{*n}*RC$, <u>and if</u> $R^{*n}*R^{\infty} \leq \underline{X} < R^{*n}*RC$, <u>then</u> $\underline{X} = R^{*n}*R^{\infty}$.

REMARKS. 1. All infinite sequences described in Lemma 12 are periodic with a period which is a power of 2. For $n > 0$, $\underline{P}_{2^n}$ is odd, by Lemma 3 and thus we find $\underline{P}_{2^n}^{\infty} = R^{*n}*R^{\infty}$ when $n \geq 1$ since $R^{*n}*R^{\infty}$ has period 2^n. However $\underline{P}_1^{\infty} = L^{\infty}$.

2. $R^{*n}*RC$ is maximal by Corollary 4 since RC is maximal.

LEMMA II.2.13. <u>When</u> $i = 2^n \cdot k$, $n \geq 0$, $k \geq 3$ <u>then</u> $\underline{P}_i = R^{*n}*RLR^{k-2}$.

<u>Proof of Theorem II.2.9</u>. This is obvious from Lemmas 10, 12, and 13.

Proof of Theorem II.2.8. By Lemmas 11 and 13 we see that the assertion of the theorem is true for pairs of numbers of the form $2^n k$, $2^{n'}k'$, with $k,k' \geq 3$, k,k' odd. Lemma 12 shows the relations between the 2^n and $2^{n'}$. The missing relation, $2^n < 2^{n'}k$ when $k \geq 3$ follows for $n' \leq n$ from $R^{*n} * R^\infty < R^{*n} * (RLR^{k-2})^\infty \leq R^{*n'} * (RLR^{k-2})^\infty$, cf. Lemma 11 and Theorem 5. When $n' > n$, we have $R^{*n} * R^\infty < R^{*n'} * R^\infty < R^{*n'} * * (RLR^{k-2})^\infty$. This proves the theorem in all cases.

Proof of Lemma II.2.10. Consider $\underline{S} = RLR^{i-2}$, $i \geq 3$, odd. By inspection $\underline{S}^\infty$ is maximal. It is periodic of period i. So we must have $\underline{P}_i \leq \underline{S}$. Since $\underline{P}_i$ is maximal, if it contains only L's or only R's, it equals L^i or R^i, which contradicts the periodicity i. Thus $\underline{P}_i$ contains R's and L's, and since it is maximal, it starts RL... . Next we claim that if $\underline{P}_i$ contains two L's then $\underline{P}_i > \underline{S}$; this observation shows $\underline{P}_i$ has one L and hence $\underline{P}_i = RLR^{i-2}$ as asserted in the lemma. To show the claim note that if the $\underline{P}_i$ did contain two L's, then $\underline{P}_i^\infty$ must contain a subsequence $\ldots RLR^{2k}L\ldots$ with $k \geq 0$, since i is odd. If q denotes the smallest such k, then by the maximality of $\underline{P}_i$ we must have $\underline{P}_i = RLR^{2q}L\ldots$, by the inequality $RLR^{2q}L\ldots > RLR^{2q+s}\ldots$, valid for $s \geq 1$. Thus if $\underline{P}_i$ contains two L's then $\underline{P}_i = RLR^{2q}L\ldots > \underline{S}$, as claimed. The proof is complete.

Proof of Lemma II.2.11. When $n = 1$ and $k \geq 7$ then

$$R * RLR^{k-2} = RLRRRL\ldots$$

and

$$RLR^{k-2} = RLRRRR\ldots .$$

The result follows in this case. The case $n = 1$, $k = 3$ or 5 is proved similarly. By Theorem 5, $R*$ is monotone and thus

$$R^{*n} * \underline{P}_k^\infty < R^{*(n-1)} * \underline{P}_k^\infty$$

implies

$$R^{*(n+1)} * \underline{P}_k^\infty < R^{*n} * \underline{P}_k^\infty \quad ,$$

so that the result follows in all cases.

Proof of Lemma II.2.12. It is obvious that $R^{\infty} < RC < R*R^{\infty}$ are consecutive among maximal sequences. Assume now that $\underline{X}$ is maximal and that

$$R^{*n}*RC \le \underline{X} < R^{*n}*(R*R^{\infty}).$$

By Theorems 5 and 7 and from $RC < R*R^{\infty}$, we find $\underline{X} = R^{*n}*\underline{B}$ with $RC \le \underline{B} < R*R^{\infty}$. Hence $\underline{B} = RC$. The other case $R^{*n}*R^{\infty} \le \underline{X} < R^{*n}*RC$ is similar.

Proof of Lemma II.2.13. We have already proved the case $n = 0$. So let $n > 0$. By Lemma 12, if $\underline{X}$ is infinite and maximal and $\underline{X} \le R^{*n}*R^{\infty}$, then the period of $\underline{X}$ is a power of 2. Thus we must have $\underline{X} > R^{*n}*R^{\infty}, \forall n$. If $\underline{X} \le \underline{S}^{\infty} \equiv R^{*n}*(RLR^{k-2})^{\infty}$, it follows thus from Theorem 5 that $\underline{X} = R^{*n}*\underline{B}$. If furthermore $\underline{X}$ is periodic with period $2^n \cdot k$, then $\underline{B}$ is periodic with period k, and $\underline{B} \le \underline{P}_k$. Since $\underline{P}_k$ is the min-max of period k and $\underline{B}$ is maximal we have $\underline{P}_k = \underline{B}$ and hence $\underline{X} = \underline{S}^{\infty}$, which is the asserted result.

Remarks and Bibliography. The calculus of itineraries is scattered in the literature and usually presented only in a circumstantial context. The most systematic account is Derrida-Gervois-Pomeau [1978], [1979] but some precursory use can be found in Metropolis-Stein-Stein [1973]. Our results up to Theorem 7 are mostly adapted from Derrida et al. but some lecture notes from Lanford [1979] were useful.

The theorem of Šarkovskii or some variant of it has been proved several times in the literature. Apart from the original reference Šarkovskii [1964], and its elaboration by Stefan [1977], there is a proof by Guckenheimer [1978], very similar to our proof, and a more recent proof (with generalizations to maps on the circle) by Block-Guckenheimer-Misiurewicz-Young [1979]. A variant of Šarkovskii's theorem is the theorem by Li-Yorke [1975], which we describe in the bibliography of the next section.

One of the aims of our presentation is to make it clear that the theorems on itineraries have a combinatorial part which is independent of the analytical part (which will be presented in the next section).

II.3 ITINERARIES AND ORBITS

The purpose of this section is to harvest the fruits of our preceding work. We shall see in fact that there is an almost one-to-one relation between itineraries and points on the line. We have already seen in Section II.1, that to every $x \in [-1,1]$, there is for fixed f an (admissible) itinerary $\underline{I}(x)$. We shall see now in which sense to every admissible, maximal itinerary $\underline{B}$ there is an x such that $\underline{B} = \underline{I}(x)$.

We will say that the extended itinerary $\underline{I}_E(x)$ of x is eventually periodic if $\mathscr{S}^k(\underline{I}_E(x))$ is periodic for sufficiently large $k \geq 0$. The period of $\mathscr{S}^k(\underline{I}_E(x))$ will then be called the eventual period of $\underline{I}_E(x)$. The relation between $\underline{I}_E(x)$ and x is described in the following lemma.

LEMMA II.3.1. *If f is unimodal then $\underline{I}_E(x)$ is eventually periodic if and only if $f^j(x)$ converges towards a periodic orbit of f as $j \to \infty$.*

LEMMA II.3.2. *If f is unimodal and $\underline{I}_E(x)$ has eventual period p, $\underline{I}_E(x) = \underline{A}\underline{B}^{\infty}$ with $|\underline{B}| = p$ then the periodic orbit of f towards which $f^j(x)$ converges has period p if $\underline{B}$ is even and period p or $2p$ if $\underline{B}$ is odd.*

All cases can occur.

Proof of Lemmas II.3.1 and II.3.2. For the proof of the "only if" part it suffices to consider the case where $\underline{I}_E(x)$ is actually periodic. If $I_{E,\ell}(x) = C$ for some ℓ, then the orbit of x is periodic and there is nothing to prove. So assume $\underline{I}_E(x)$ does not contain C, and is periodic of period p. Then we define sets $J_0,\ldots,J_{p-1}$ as follows. For every r, $0 \leq r < p$, J_r is the smallest closed subinterval of $[-1,1]$ containing all the points $f^{r+pn}(x)$, where $n > 0$. Then the interior of J_r does not contain 0 and hence $f|_{J_r}$ is a homeomorphism into $J_{r+1 \pmod p}$. Thus f^p is a homeomorphism of J_0 into itself. We distinguish two cases.

CASE 1. $f^p|_{J_0}$ preserves orientation. Then if $x \leq f^p(x)$ it follows that

$$x \leq f^p(x) \leq f^{2p}(x) \leq \ldots$$

and hence this subsequence converges to a limit in J_0. This limit belongs to a periodic orbit whose period q divides p. If $x \geq f^p(x)$ the inequalities

$$x \geq f^p(x) \geq f^{2p}(x) \geq \ldots$$

will lead to the same conclusion.

CASE 2. $f^p|_{J_0}$ reverses orientation. Then $f^{2p}|_{J_0}$ preserves orientation and we proceed as before.

If $f^n(x)$ converges to a periodic orbit of period q which does not contain 0, then the "if" part of the lemma is obvious by continuity, and $\underline{I}_E(x)$ must be periodic with a period p dividing q.

If the periodic orbit does contain zero, then one can see from the continuity and the piecewise monotonicity of f that if $f^{r+qn}(x)$ converges to zero for $n \to \infty$ then all $f^{r+qn}(x)$ must be on the same side of zero for sufficiently large n. Hence the assertion follows in this case. Since q divides $2p$ or p and p divides q, we find the conclusion of Lemma 2. Q.E.D.

We illustrate the occurrence of both alternatives. When $f(x) = 1 - 1.75x^2$, the extended itinerary of 0 is

$$\text{CRLRRLRR}\ldots = \text{C(RLR)}^\infty$$

and the corresponding periodic orbit is

$$0.9083, \quad -0.7440, \quad 0.0314.$$

When $f(x) = 1 - 1.77x^2$, the extended itinerary of 0 is

$$CRLLRLLRLL\ldots = C(RLL)^{\infty},$$

but the periodic orbit is

$$0.996, \quad -0.756, \quad -0.012, \quad 0.9997, \quad -0.769, \quad -0.047.$$

When $f(x) = 1 - 1.75487\ldots x^2$, the extended itinerary of 0 is $(CRL)^{\infty}$.

So we find all cases as asserted. For a more complete discussion, see Section II.6.

NOTE. The preceding concepts are easily extended to functions which are piecewise strictly monotone. Let f be strictly monotone on the ℓ intervals $[c_0,c_1]$, $[c_1,c_2],\ldots$ $[c_{\ell-1},c_\ell]$ with $c_0=-1$, $c_\ell=1$. If $f^k(x) \in (c_{j-1},c_j)$, then $I_k(x)$ is defined to be L_j and if $f^k(x) = c_j$ then $I_k(x) = C_j$. The itinerary of x is thus a sequence made of symbols $L_1,\ldots,L_\ell$ and of $C_0,\ldots,C_\ell$. The above Lemma 1 and its proof generalizes to this case.

The following lemmas will be useful for associating periodic points to periodic itineraries.

LEMMA II.3.3. _Let_ f _be unimodal and suppose_ x _is periodic with period_ $2p$ _but_ $\underline{I}(x)$ _is periodic with period_ p. _Then there is an_ x' _which is periodic with period_ p, _and_ $\underline{I}(x') = \underline{I}(x)$.

Proof. Suppose $\underline{I}(x) = \underline{D}^{\infty}$ with $|\underline{D}| = p$. Define J as the interval with endpoints x and $f^p(x)$. Then by the same arguments as in the proof of Lemmas 1,2, f^p is an orientation reversing homeomorphism of J onto itself, since $\underline{D}$ is odd by Lemmas 1, 2. Thus f^p has a fixed point in J, by continuity.

LEMMA II.3.4. *Let f be unimodal and suppose there is a y such that $\underline{I}(y)$ has period p. Then there is an x' which has period p. The itinerary of x' may turn out to be different from that of y.*

Proof. Write $\underline{I}(y) = \underline{D}^{\infty}$, $|\underline{D}| = p$. If $\underline{D}$ is even, the assertion follows from Lemma 2. If $\underline{D}$ is odd, there is by Lemma 2 a periodic point z of period p or of period $2p$. It the period is p we are finished. If the period is $2p$ and $\underline{I}(z)$ is infinite, we must have $\underline{I}(z) = \underline{I}(x)$ and the result follows from Lemma 3. In the remaining case we may assume without loss of generality that $z = 1$, $\underline{D}$ maximal and $\underline{I}(1) = \underline{B}C$, $|\underline{B}| = 2p-1$. We then have $B_0 = B_p = D_0$, $B_1 = B_{p+1} = D_1, \ldots, B_{p-2} = B_{2p-2} = D_{p-2}$, $B_{p-1} = D_{p-1}$. With these precautions, we may repeat the argument of Lemma 3, taking $J = [f^p(1), 1]$.

The main input for our analysis of whether the map $x \to \underline{I}(x)$ is "onto" are the following topological results.

PROPOSITION II.3.5. *Let f be a unimodal function. Assume $\underline{I}(1)$ is an infinite sequence. Let $\underline{A}$ be an admissible sequence satisfying $\mathscr{S}^k\underline{A} < \underline{I}(1)$ for all $k \geq 0$. Then the sets*

$$L_{\underline{A}} \equiv \{x: x \in (-1,1) \quad \text{and} \quad \underline{I}(x) < \underline{A}\}$$

and

$$R_{\underline{A}} \equiv \{x: x \in (-1,1) \quad \text{and} \quad \underline{I}(x) > A\}$$

are open.

PROPOSITION II.3.6. *Let f be a unimodal function. Assume $\underline{I}(1) = \underline{D}C$. Let $\underline{A}$ be an admissible sequence. If $\underline{D}$*

<u>is even and</u> $\mathcal{S}^k\underline{A} < (\underline{D}L)^\infty$ <u>for all</u> $k \geq 0$ <u>or if</u> $\underline{D}$ <u>is odd and</u> $\mathcal{S}^k\underline{A} < (\underline{D}R)^\infty$ <u>for all</u> $k \geq 0$ <u>then</u> $L_{\underline{A}}$ <u>and</u> $R_{\underline{A}}$ <u>are open</u>.

If a sequence $\underline{A}$ satisfies the conditions of Propositions 5 or 6, i.e., $\underline{A}$ is admissible and for all $k \geq 0$,

$\mathcal{S}^k\underline{A} < \underline{I}(1)$ if $\underline{I}(1)$ is infinite,

$\mathcal{S}^k\underline{A} < (\underline{D}L)^\infty$ if $I(1) = \underline{D}C$ and $\underline{D}$ is even,

$\mathcal{S}^k\underline{A} < (\underline{D}R)^\infty$ if $\underline{I}(1) = \underline{D}C$ and $\underline{D}$ is odd,

we shall say that $\underline{A}$ is <u>dominated</u> by $\underline{I}(1)$, and we shall use the notation $\underline{A} \ll \underline{I}(1)$.

This definition is reasonable since $\underline{A}$ should be thought of as the itinerary of a point in $[-1,1]$. One might be tempted to replace the strict inequalities in the definition of $\ll$ by $\leq$. That the results do not hold anymore will be shown after the proof of Theorem 8.

<u>Proof of Proposition II.3.5</u>. Let us consider $R_{\underline{A}}$. Assume $\underline{I}(y) > \underline{A}$ for some $y \in (-1,1)$. We show that there is some neighborhood $\mathcal{U}$ of y so that the inequality $\underline{I}(x) > \underline{A}$ holds for $x \in \mathcal{U}$.

1. If $\underline{I}(y)$ is an infinite sequence, then we argue as follows. Let n be the first index for which $I_n(y) \neq A_n$. Since $\underline{I}(y)$ is infinite, $I_n(y)$ equals R or L. Thus by the continuity of f and hence of f^n we can preserve the equalities $I_j(x) = A_j$ when $j < n$ and the inequality $I_n(x) \neq A_n$ for x sufficiently near to y.

2. If $\underline{I}(y)$ is a finite sequence and if n is defined as before, the same argument can be applied if $I_n(y) \neq C$.

3. There remains thus the case $I_n(y) = C$. Then we have $\underline{I}(y) = \underline{B}C$ and $\underline{A} = \underline{B}L\hat{\underline{A}}$ if $\underline{B}$ is even, or $\underline{A} = \underline{B}R\hat{\underline{A}}$ if $\underline{B}$ is

odd. We consider only the proof for the first alternative. The other case is handled analogously. The extended itinerary of y is $\underline{I}_E(y) = \underline{B}C\underline{I}(1)$. Since $\underline{A} \ll \underline{I}(1)$, we must have $\hat{\underline{A}} < \underline{I}(1)$. Let m denote the first index for which $\hat{A}_m \neq I_m(1)$. By assumption, $I_m(1) \neq C$. There is thus an $x_m < 1$ such that for $z \in (x_m, 1]$ we have $I_j(z) = \hat{A}_j$ for $j < m$ and $I_m(z) \neq \hat{A}_m$. But 1 is the image of zero (i.e., C). Thus some neighborhood $(-w,w)$ of zero is mapped into $(x_m,1]$. For y' sufficiently near to y, we will thus find $f^n(y') \in (-w',w')$, with w' as small as we please, and in particular $0 < w' \leq w$. Thus we can achieve $f^{n+1}(y') \in (x_m,1]$ and sign $f^j(y') = \text{sign } f^j(y)$ for $j = 0,1,\ldots,n-1$. Thus for such $y' \neq y$, $\underline{I}_E(y')$ is of the form

$$\underline{I}_E(y') = \underline{B}RI_0(1)\ldots I_m(1)\underline{D}$$

or

$$\underline{I}_E(y') = \underline{B}LI_0(1)\ldots I_m(1)\underline{D}$$

while

$$\underline{A} = \underline{B}LI_0(1)\ldots I_{m-1}(1)I'_m(1)\underline{E}$$

with $I'_m(1) \neq I_m(1)$. Thus $\underline{I}_E(y') > \underline{A}$, since $\underline{B}$ is even and $\hat{\underline{A}} < \underline{I}(1)$. This completes the proof that $R_{\underline{A}}$ is open. The case of $L_{\underline{A}}$ is similar.

Proof of Proposition II.3.6. The proof is very similar to the preceding one. The Cases 1 and 2 are identical since $\underline{I}(1)$ does not enter the argument. The Case 3 is handled by the next lemma.

LEMMA II.3.7. *Let f be unimodal and assume $\underline{I}(1) = \underline{D}C$. Given $s \geq 1$, for $x \neq 0$ sufficiently near to 0, $\underline{I}(f(x)) = (\underline{D}L)^s\ldots$ if $\underline{D}$ is even and $(\underline{D}R)^s\ldots$ if $\underline{D}$ is odd.*

3'. There remains the case $I_n(y) = C$. Then we have $\underline{I}(y) = \underline{B}C$ and $\underline{A} = \underline{B}L\hat{\underline{A}}$ if $\underline{B}$ is even, or $\underline{A} = \underline{B}R\hat{\underline{A}}$ if $\underline{B}$ is odd. We consider only the proof for the first alternative. The extended itinerary of y is $\underline{I}_E(y) = \underline{B}C\underline{I}(1)^\infty$. By the arguments used in the proof of Proposition 5, for y' sufficiently near to y we have that $f^n(y)$ is very near to zero and hence, for given s, by Lemma 7, $\underline{I}_E(y') = \underline{B}\underline{L}(\underline{D}\underline{L})^s\underline{X}$ or $\underline{B}\underline{R}(\underline{D}\underline{L})^s\underline{X}$ when $\underline{D}$ is even. (And similarly $\ldots(\underline{D}\underline{R})^s\ldots$ when $\underline{D}$ is odd. We shall not mention this case anymore.) Since $\underline{A} \ll \underline{I}(1)$ we have $\hat{\underline{A}} < (\underline{D}\underline{L})^\infty$ i.e., $\hat{\underline{A}} = (\underline{D}\underline{L})^{s-1}\underline{E}$ with $\underline{E} < \underline{D}\underline{L}\underline{X}$, $\underline{E} \neq \underline{D}\ldots$ for some s. Therefore we have $\underline{I}_E(y') > \underline{A} = \underline{B}\underline{L}\hat{\underline{A}}$. The proof that $R_{\underline{A}}$ is open is complete.

<u>Proof of Lemma II.3.7</u>. Let p be the period of 0, $f^p(0) = 0$, i.e., $p = |\underline{D}| + 1$. Consider now f^{p-1}. According to Remark II.1.4, there is an interval $(w_0,1]$ on which f^{p-1} is strictly monotone. The itinerary of 1 starts $\underline{D}C\ldots$. Thus the itinerary of $x \in (w_0,1)$ starts $\underline{D}L\ldots$ when $\underline{D}$ is even, since then f^{p-1} preserves orientation. It follows that f^p maps $(w,1]$ into $(w_0,1]$ provided w is sufficiently near to 1. Furthermore, given $w_1 < 1$ (near to 1), we can always achieve $f^p(w,1] \subset (w_1,1]$. Iterating s times in this fashion, we can find a $w < 1$ such that $f^{kp}(w,1] \subset (w_1,1]$ for $k = 0,1,2,\ldots,s$. Thus the itinerary of $y \in (w,1)$ starts $(\underline{D}L)^s\ldots$. By the continuity of f, there is now a neighborhood $\mathscr{U}$ of 0, such that $f\mathscr{U} \subset (w,1]$. This proves the assertion of the lemma when $\underline{D}$ is even. The argument is analogous when $\underline{D}$ is odd.

We can now state the main result of this section.

THEOREM II.3.8. <u>Let</u> f <u>be unimodal, and assume</u> $\underline{A}$ <u>is an admissible sequence satisfying</u>

$$\underline{I}(-1) \leq \underline{A} \ll \underline{I}(1) \quad .$$

<u>Then there is an</u> $x \in [-1,1]$ <u>such that</u> $\underline{I}(x) = \underline{A}$.

REMARK. The condition of the theorem implies $\mathscr{S}^k\underline{A} \geq \underline{I}(-1)$ for all $k \geq 0$.

<u>Proof</u>. In view of our previous work, the proof is now very easy. By Propositions 5 and 6, $L_{\underline{A}}$ and $R_{\underline{A}}$ are open. Thus

$$L^{\perp}_{\underline{A}} \equiv \{x: \underline{I}(x) \geq \underline{A} \quad \text{and} \quad x \in [-1,1]\}$$

and

$$R^{\perp}_{\underline{A}} \equiv \{x: \underline{I}(x) \leq \underline{A} \quad \text{and} \quad x \in [-1,1]\}$$

are closed. Since $\underline{I}(-1) \leq \underline{A} \leq \underline{I}(1)$ neither of $L^{\perp}_{\underline{A}}$ and $R^{\perp}_{\underline{A}}$ is empty. Furthermore, $L^{\perp}_{\underline{A}} \cup R^{\perp}_{\underline{A}} = [-1,1]$. Since $[-1,1]$ is connected, $L^{\perp}_{\underline{A}} \cap R^{\perp}_{\underline{A}}$ is not empty. Q.E.D.

Theorem 8 tells us that every "reasonable" sequence sandwiched between $\underline{I}(-1)$ and $\underline{I}(1)$ does occur as the itinerary of at least one point $x \in [-1,1]$. One may now ask whether our restrictions in the definition of "reasonable" (i.e., dominated) sequences are necessary.

We show that in the definition of $<<$ one cannot allow $\leq$ signs. If we did, in the first case, $\underline{I}(1)$ infinite, we would have to find a y with $\underline{I}(y) = L\underline{I}(1)$. We shall give now an example of a function for which this is not possible. The example is based on a general observation about the piecewise linear maps $f_{\mu}(x) = 1 - \mu|x|$ when $\mu \in (1,2]$. For such maps, $\underline{I}(x) \neq \underline{I}(x')$ if $x \neq x'$. This is so because if $x \neq x'$ are both on the same side of 0, then $|f^k(x) - f^k(x')| = \mu^k|x-x'|$, as long as $f^j(x)$ and $f^j(x')$ are on the same side of 0 for $0 \leq j < k$. Since $\mu > 1$, k cannot be infinite. We apply this to the function $f(x) = 1 - \mu|x|$, with $\mu = \sqrt{2}$. Then $f(1) = 1 - \sqrt{2}$, $f^j(1) = \sqrt{2}-1$ for $j \geq 2$, i.e., $\underline{I}(1) = RLR^{\infty}$. By the argument just given, if $\underline{I}(y) = LRLR^{\infty}$ then $\underline{I}(f(y)) = \underline{I}(1)$, but since $y \neq 0$, we have $f(y) \neq 1$, a contradiction. Thus the first inequality $\mathscr{S}^k\underline{A} < \underline{I}(1)$ in the definition of $<<$ is justified. Concerning the other inequality, let $f(x) = 1 - \mu|x|$ with $\mu = (1 + \sqrt{5})/2$. Then $f(1) = 1 - \mu$, $f^2(1) = 1 + \mu - \mu^2 = 0$, so $\underline{I}(1) = RLC$. We show there is no y with $\underline{I}(y) = (RLR)^{\infty} \equiv (\underline{D}R)^{\infty}$. Again by the above observation on piecewise linear maps, $f^k(y)$ and $f^k(1)$ cannot forever lie on the same side of 0 (0 included!). So there cannot be a y with $\underline{I}(y) = (\underline{D}R)^{\infty}$. This justifies the other inequalities in the definition of $<<$.

When $\underline{I}(1)$ is finite, $\underline{I}(1) = \underline{B}C$ with, say, $\underline{B}$ even, then if f is $\mathscr{C}^1$-unimodal, there exists always a point y with $\underline{I}(y) = (\underline{B}L)^\infty$. In fact one has

LEMMA II.3.9. Let f be $\mathscr{C}^1$-unimodal and $\underline{I}(1) = \underline{B}C$. For all $x < 1$ sufficiently near to 1, one has $\underline{I}(x) = (\underline{B}L)^\infty$ if $\underline{B}$ is even, $\underline{I}(x) = (\underline{B}R)^\infty$ if $\underline{B}$ is odd.

Proof. Lemma 9 is the $\mathscr{C}^1$ variant of Lemma 7. Instead of the iterates of 1 we consider the iterates of 0. Assume $|\underline{B}C| = p$. Then $f^p(0) = 0$, and hence $|Df(z)| < \varepsilon$ for $|z|$ sufficiently small, since f is $\mathscr{C}^1$-unimodal. When $\underline{B}$ is even, then $Df^p(z) < 0$ when $z > 0$ is small, and $Df^p(z) > 0$ when $z < 0$ is small. Thus $f^p(z) = \int_0^z Df^p(\zeta)\, d\zeta < 0$ and if $|z| \neq 0$ is small then $0 < |f^p(z)| < \varepsilon |z|$ is also small. This proves the assertion when $\underline{B}$ is even. Similarly for odd $\underline{B}$.

Theorem 8 tells us that every "reasonable" sequence sandwiched between $\underline{I}(-1)$ and $\underline{I}(1)$ actually occurs as the itinerary of at least one point. We specialize the situation by considering periodic sequences. Then one has the following result, (which is weaker than Theorem 8).

THEOREM II.3.10. Šarkovkii's Theorem. Reconsider the ordering of integers

$$
\begin{aligned}
&3 > 5 > 7 > 9 > \ldots \\
&> 2\cdot 9 > 2\cdot 5 > 2\cdot 7 > 2\cdot 9 > \ldots \\
&\ldots \\
&> 2^n\cdot 3 > 2^n\cdot 5 > \ldots \\
&\ldots \\
&> \ldots > 2^m > \ldots > 8 > 4 > 2 > 1.
\end{aligned}
$$

If f is unimodal and has a point with period p then it has a point with period q for every $q < p$ in the sense of the above ordering. It seems that it must be explicitly stated that the theorem does not assert the existence of stable periodic orbits. We shall see later that many unimodal maps have at most one stable periodic orbit.

Proof. Assume that x is a periodic point of period $p \neq 2^n$, and suppose $\underline{I}(x)$ is infinite. Without loss of generality we may assume $\underline{I}(x)$ is maximal. By Lemma 2, the period of $\underline{I}(x)$ is p or $p/2$ and by the pre-Šarkovskii theorem this implies $\underline{I}(x) \geq \underline{P}_p^\infty$. By Lemma II.1.3, $\underline{I}(x) \leq \underline{I}(1)$.

Thus $\underline{P}_p^\infty \leq \underline{I}(1)$. If $q < p$ in the sense of the Šarkovskii ordering, then $\underline{P}_q^\infty < \underline{P}_p^\infty$ by the pre-Šarkovskii theorem, Theorem II.2.8. Since $\underline{P}_q^\infty$ is maximal, this implies $\underline{P}_q^\infty << \underline{I}(1)$.

Assume next x is a periodic point of period $p \neq 2^n$ and $\underline{I}(x)$ is finite. Thus we may assume $x = 1$. Write $p = 2^n k$, $n \geq 0$, $k \geq 3$, odd, and write $\underline{I}(1) = \underline{D}C$, with $|\underline{D}C| = p$. Assume for definiteness that $\underline{D}$ is even. If the period of $(\underline{D}L)^\infty$ equals p then we must have $\underline{P}_p^\infty \leq (\underline{D}L)^\infty$, and we can argue as in the case of infinite $\underline{I}(x)$. If the period of $(\underline{D}L)^\infty$ equals s, where s divides p, then $(\underline{D}L)^\infty \geq \underline{P}_s^\infty > \underline{P}_p^\infty$ unless $s = 2^n$, by the Šarkovskii ordering. But the only remaining possibility, $s = 2^n$ cannot occur: By Lemma II.2.12, $\underline{D}L = \underline{P}_s^k$; but $\underline{P}_s$ is odd, (by II.2.3), $\underline{D}L$ is even and k is odd (similarly $\underline{D}R$ is even if $\underline{D}$ is odd).

In summary, we have shown so far that if x has period $p \neq 2^n$ then $\underline{P}_q^\infty << \underline{I}(1)$ for all $q < p$ in the Šarkovskii ordering. Every unimodal map must have a fixed point (>0) because $f(0) = 1$, $f(1) < 1$. So let us not consider anymore the somewhat singular case of $\underline{P}_1^\infty = L^\infty$, by assuming $q > 1$ and $q < p$ in the Šarkovskii ordering. Now $\underline{I}(-1)$ starts $L\ldots$ and every $\underline{P}_q$, $q > 1$ starts $R\ldots$, so that $\underline{P}_q^\infty > \underline{I}(-1)$. There is thus by Theorem 8 a $y \in [-1,1]$ with $\underline{I}(y) = \underline{P}_q^\infty$. It follows from Lemma 4 that there is a z with period q.

We now discuss the case when x has period $p = 2^n$. If $\underline{I}(x)$ is infinite (and maximal) and of period 2^n then

$\underline{I}(x) \geq \underline{P}^{\infty}_{2^n} >> \underline{P}^{\infty}_{2^{n-1}}$. The argument proceeds as before. If $\underline{I}(x)$ has period 2^{n-1} we can find by Lemma 4 a y with period 2^{n-1}. For the other periods the proof proceeds as before. Finally we remain with the somewhat special case $p = 2^n$, $\underline{I}(x)$ finite, and we assume $x = 1$. If $\underline{I}(1) > R^{*(n-1)}*RC$ then $\underline{I}(1) >> R^{*(n-1)}*R^{\infty}$ and we proceed as before. If $\underline{I}(1) = R^{*(n-1)}*RC$, then by a variant of the proof of Lemma 4 one exhibits a point with period 2^{n-1}.

Remarks and Bibliography. The paper of Milnor-Thurston [1977] was the first to raise the question of identifying itineraries and orbits but it answers this only in part. They deal mostly with periodic itineraries. The ideas of Proposition 5 can be found in Guckenheimer [1979] and Lanford [1979]. The way we have developed the theory, Sarkovskii's theorem [Šarkovskii, 1964; Štefan, 1977] is presented in such a fashion that it appears in its proper place as a corollary of the general results of Proposition 5. For a systematic exposition using characteristic polynomials see Jonker-Rand [1980]. A similar result could be presented in the same way: The "Period 3 implies chaos" of Li-Yorke [1975].

THEOREM. Let f be a unimodal map with a periodic point of period 3. Then there is an uncountable set S of points and an $\varepsilon > 0$ such that for every $x, y \in S$, $x \neq y$

$$\limsup_{n\to\infty} |f^n(x) - f^n(y)| \geq \varepsilon$$

and

$$\liminf_{n\to\infty} |f^n(x) - f^n(y)| = 0 \quad .$$

(Sketch of a proof in the spirit of our method: If f has period 3, then $\underline{I}(1) \geq (RLR)^{\infty}$. Consider now itineraries of the form $RLR^{p_1}LR^{p_2}LR^{p_3}\ldots$ with $p_i \to \infty$ as $i \to \infty$. We call two itineraries equivalent if they have the same ends. Pick one representative from each equivalence class and let S be the set of corresponding points according to Theorem 8.)

II.4 NEGATIVE SCHWARZIAN DERIVATIVE

Let f be $\mathscr{C}^1$-unimodal, and let P be a periodic orbit of period p for f. We call P a stable periodic orbit if for $x \in P$, $|Df^p(x)| \leq 1$. By the chain rule of differentiation, $Df^p(x)$ takes the same value for all $x \in P$, so that the definition makes sense. The importance of stable periodic orbits for dynamical systems comes from the following observation. If P is a stable periodic orbit for f, (of period p), then there is some neighborhood U of $x \in P$ such that $\lim_{n\to\infty} f^{np}(y) = x$ for all $y \in U$ except possibly if $|Df^p(x)| = 1$, see below. Thus, in the case of a stable periodic orbit, many points have similar behavior as $n \to \infty$. A periodic orbit is called superstable if $0 \in P$ (i.e., iff $Df^p(x) = 0$ for $x \in P$).

We now ask how many stable periodic orbits a unimodal map can have. This question was first asked by Julia, in 1918. He showed that for certain unimodal maps which are restrictions to $[-1,1]$ of analytic functions, there can be at most one stable periodic orbit. In particular, his theory applies to $f(x) = 1 - \mu x^2$, $0 < \mu \leq 2$. But a real breakthrough has been made by Singer [1978], who isolated the condition of negative Schwarzian derivative as the simplifying feature. We now present his theory.

Assume that f is $\mathscr{C}^3$. The Schwarzian derivative of f at x, denoted $Sf(x)$ is defined by

$$Sf(x) = \frac{f'''(x)}{f'(x)} - \frac{3}{2}\left(\frac{f''(x)}{f'(x)}\right)^2 .$$

We shall call f S-unimodal if

S1. f is $\mathscr{C}^1$-unimodal.

S2. f is $\mathscr{C}^3$.

S3. $Sf(x) < 0$ for all $x \in [-1,1]$, at $x = 0$ we allow the value $-\infty$ for $Sf(x)$.

S4. f maps $J(f) = [f(1),1]$ onto itself.

S5. $f''(0) < 0$.

Recall that the definition of $\mathscr{C}^1$-unimodal includes the condition that the only critical point of f is $x = 0$. The condition S4 is included mainly for convenience of the statements of results. It is satisfied for all even unimodal functions. Note that the property of being S-unimodal is stable under small $\mathscr{C}^3$-perturbations.

The condition S5 is not needed in this section, but only starting from Section 5. To simplify the statements, we nevertheless say "S-unimodal f" in this section instead of "f satisfying S1-S4". The main result of this section is the

THEOREM II.4.1. *If f satisfies S1, S2, S3, then every stable periodic orbit attracts at least one of the points -1, 0, 1 (i.e., the end points of the interval, or the critical point).*

Two important corollaries follow from this.

COROLLARY II.4.2. *If f is S-unimodal, then it has at most one stable periodic orbit, plus possibly a stable fixed point in the interval $[-1,f(1))$. If 0 is not attracted to a stable periodic orbit then f has no stable periodic orbit in $[f(1),1]$.*

COROLLARY II.4.3. *There exist S-unimodal functions without a stable periodic orbit.*

We shall prove Theorem 1 and Corollary 2 later.

<u>Proof of Corollary II.4.3</u>. The following classical example has been given by Ulam and V. Neumann [1947]: $f(x) = 1 - 2x^2$. It is easy to check that f is S-unimodal. The orbit of 0 is $0, 1, -1, -1, \ldots$. By Theorem 1 there can be at most one stable periodic orbit, and it must attract one of the points -1, 0 or 1. But $f(-1) = -1$ is a fixed point and attracts these three points. Hence they can be attracted by nothing else. The only remaining possibility is that -1 itself is a stable fixed point. But this is not the case since $f'(-1) = 4$.

Of course we can invent other maps, for which some iterate of 0 falls into an unstable periodic orbit, not just into an unstable fixed point. Then Corollary 2 implies that f has no stable periodic orbit in $[f(1),1]$. We shall see in Section 8 that maps satisfying the above hypotheses have ergodic properties.

COROLLARY II.4.4. <u>Assume</u> f <u>is</u> S-<u>unimodal</u>. <u>If</u> $f'(-1)>1$ <u>then the alternative of a fixed point in</u> $[-1,f(1))$ <u>cannot occur</u>.

This corollary will also follow from the proof of Theorem 1. The whole discussion about the additional fixed point in $[-1,f(1))$ is thus slightly artificial, since by Corollary 2, if -1 and 0 are attracted to two different orbits, then -1 must be attracted to a stable fixed point in $[-1,f(1))$. Thus, the whole problem could be eliminated by talking about maps on $[f(1),1]$ rather than on $[-1,1]$, which we have chosen mostly for notational simplicity.

Before presenting the proof of Theorem 1, we want to insist that the condition of negative Schwarzian derivative cannot be totally disposed of, if we want the conclusion of Theorem 1 to hold. The fact that f is unimodal is in itself not sufficient to ensure the existence of only one stable periodic orbit in $[f(1),1)$. Even if f is concave, this is not sufficient as shown in the following example. We take

$f(x) = 1 - 1/2(x^2 + x^{14})$. This is a concave unimodal function with superstable period $\{0,1\}$ of period 2 and a stable fixed point at $x_0 = 0.72861...$ with $f'(x_0) = -0.84279...$. This stable fixed point does not attract any of the three points 0, 1, -1. See Figure below.

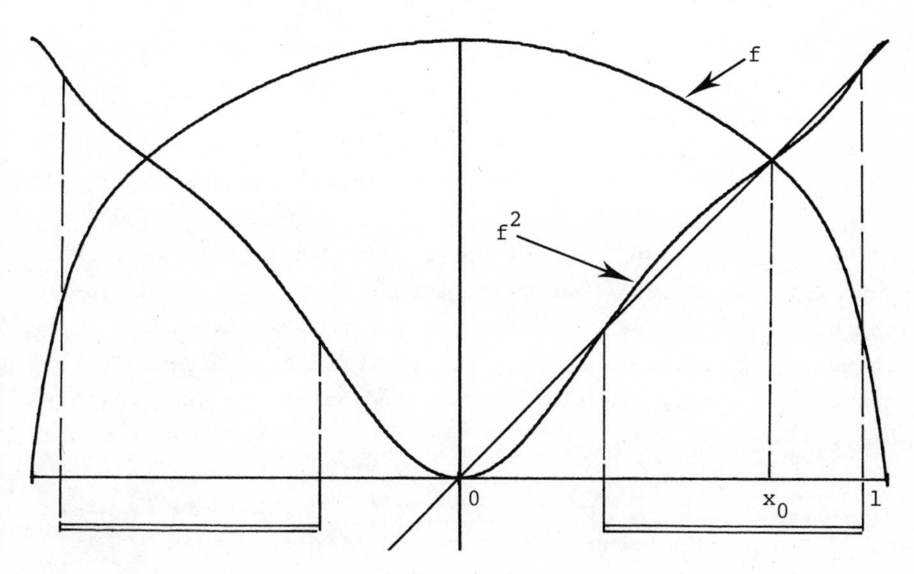

Figure II.4. $f(x) = 1 - 0.5(x^2 + x^{14})$. x_0 is a stable fixed point. Solid bars indicate stable manifold of x_0.

We now present Singer's proof of Theorem 1 and indicate then where the conditions can be relaxed.

1. Let $f, g \in \mathscr{C}^3$. Then $S(f \circ g)(x) = (Sf)(g(x))g'(x)^2 + Sg(x)$. This follows by a direct computation.

2. It follows that if $f \in \mathscr{C}^3$ and $Sf(x) < 0$ for all $x \in [-1,1]$, then $S(f^n)(x) < 0$ for all $x \in [-1,1]$.

3. If $f \in \mathscr{C}^3$ and $Sf(x) < 0$ for all x, then $|f'|$ has no positive local minimum, in $(-1,1)$.

Proof. If f' has an extremum at y, $y \in (-1,1)$ then $f''(y) = 0$ and $Sf(y) < 0$ implies that $f'''(y)$ and $f'(y)$ have opposite signs. In particular, this shows that f cannot be locally conjugate to $x + x^3$, and if $|f'(y)| = 1$, then on one side of y, $|f'| < 1$, locally. Thus every fixed point $y \neq \pm 1$ with $|f'(y)| = 1$ is stable from at least one side, cf. also 7).

4. Let now $f \in \mathscr{C}^3$ have finitely many critical points and satisfy $Sf(x) < 0$ for all $x \in [-1,1]$. Then f has only finitely many points of period n for every integer $n \geq 1$.

Proof. Let $g = f^n$ and suppose $g(x) = x$ for infinitely many x. Then by the mean value theorem, $g'(x) = 1$ for infinitely many x. By 2 and 3, $|g'|$ has no positive local minimum, and therefore it must vanish infinitely often. This contradicts the hypothesis that f, and hence g has finitely many critical points.

5. If $a < b < c$ are consecutive fixed points of $g = f^n$, and if $[a,c]$ contains no critical point of g, then $g'(b) > 1$.

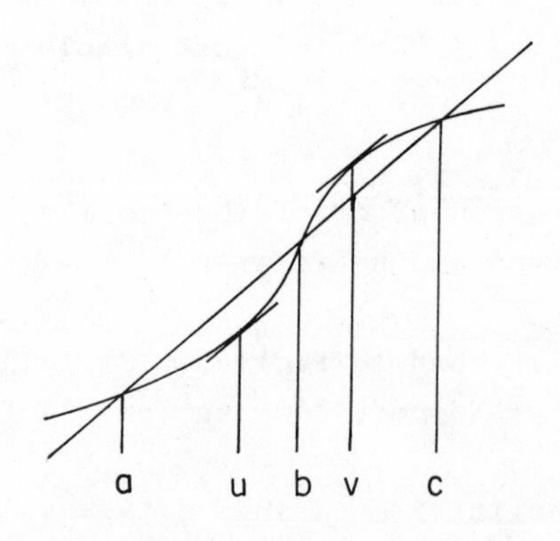

Figure II.5.

Proof. By the mean value theorem, there are u, v, $a < u < b < v < c$ such that $g'(u) = g'(v) = 1$. If $g'(x) > 0$ on $[a,c]$, then $g'(b) > 1$, by 3.

6. Assume $x \in (-1,1)$ is a stable fixed point for $g = f^n$, and assume $|g'(x)| < 1$. Define the <u>stable manifold</u> of x as the set of points y for which $g^m(y) \to x$ as $m \to \infty$, and the <u>semilocal stable manifold</u> of x as the connected component of the stable manifold of x, which contains x. If x is in the interior of $[-1,1]$ this is an open interval (r,s), or it is of the form $[-1,s)$ or $(r,1]$ (or $[-1,1]$, a trivial variant). Let us consider the case (r,s) first. g maps the semilocal stable manifold of x into itself but not r or s into it, since they are not in the stable manifold. The only possibilities are one of the three below

(i) $g(r) = r$ and $g(s) = s$

(ii) $g(r) = s$ and $g(s) = r$

(iii) $g(r) = g(s)$ $(= r$ or $s)$.

Case (i) is eliminated by 5, with $r = a$, $x = b$, $s = c$ and Case (ii) is eliminated similarly by 5 considering g^2 instead of g. Thus $g(r) = g(s)$. Thus by Rolle's theorem, g has a critical point in (r,s), which is attracted to x.

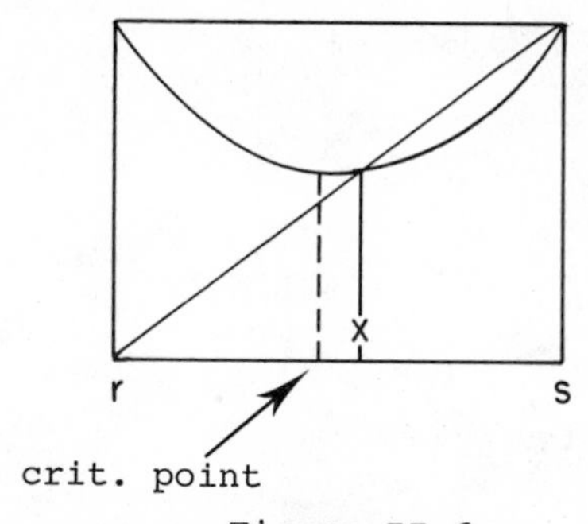

Figure II.6.

But if g has a critical point p in (r,s) and $f^n = g$, then a critical point of f is mapped into p.

Let us now consider the case when the semilocal stable manifold is $[-1,s)$. Then by definition, -1 is attracted to x, and we argue similarly for $(r,1]$. This shows Theorem 1 in all cases when $|g'(x)| < 1$, except if $x = +1$ or -1. But then there is nothing to prove.

7. Let $g = f^n$ and $g(x) = x$, $|g'(x)| = 1$. By considering g^2 instead of g if necessary we may assume without loss of generality that $g'(x) = 1$. If $x = +1$ or -1 there is nothing to prove. If $x \in (-1,1)$, there is, by 4, a neighborhood (r,s) of x containing no other fixed points of g. Either $g(y) > y$ for all $y \in (r,x)$ or $g(y) < y$ for all $y \in (x,s)$ otherwise $g'(y) > 1$ would have solutions in (r,s) on both sides of x and g' would have a positive local minimum. So assume for definiteness $g(y) > y$ for $y \in (r,x)$. At the minimal value d of y for which $g(y) \geq y$, we have $g(d) = d$ (or $d = -1$ leading to the attraction of the end point). At this point $g'(d) \geq 1$, and since there is a point w in (d,x) at which $g'(w) = 1$, we find the assertion from 3, cf. Figure II.7.

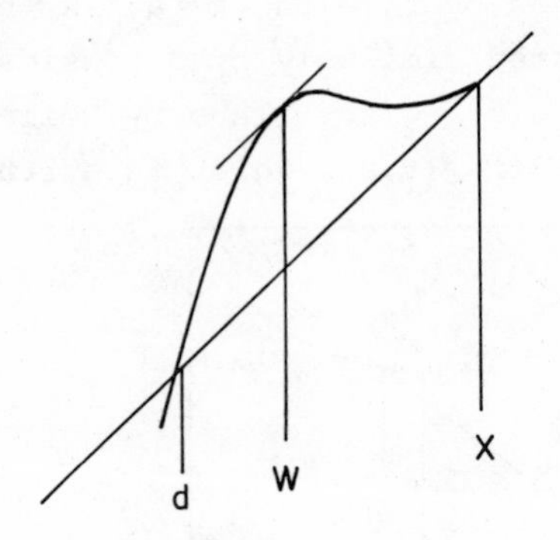

Figure II.7.

This completes the proof of Theorem 1.

Before entering the proof of Corollaries 2 and 4, we give an example showing the occurrence of attraction of an end point.

Note that the proof of Theorem 1 did not use the fact that f is $\mathscr{C}^3$ at $x = 0$, but only $\mathscr{C}^1$ is needed. This will allow us to show an easier example. Define $f(x) = \beta(\exp(-\alpha x^2) - 1) + 1$ for $x \le 0$ and $f(x) = 1 - x^2$ for $x \ge 0$. When $\beta > 0$ and $\alpha \in \mathbb{R}$, $\alpha \ne 0$ it is easy to check that $(Sf)(x) < 0$ for $x \ne 0$. By varying α, β suitably, we can illustrate the different possibilities of Theorem 1. In all cases 0 is attracted by the superstable period $\{0,1\}$. When $\beta(\exp(-\alpha)-1) = -2$ then $f(-1) = -1$ and $f'(-1) = 2\alpha\beta e^{-\alpha}$. In particular we have the values

α	β	$f(-1)$	$f'(-1)$
7	2.0018...	-1	< 1
2.3367...	2.214...	-1	= 1

But we can also achieve to have a fixed point x in $[-1, f(1))$ with $f'(x) = 1$ or $f'(x) < 1$, for $\alpha = 7$ and $\beta = 1.738...$ and $\beta = 1.9$ respectively.

All these cases are shown in Figure II.8. In particular, we see that negative Schwarzian derivative implies $|f'|^{-1/2}$ is convex, but that f is not necessarily concave.

<u>Proof of Corollary II.4.2</u>. Since $f(0) = 1$, the points 1 and 0 are attracted to the same periodic orbit. If a periodic orbit has one point in $J(f) = [f(1),1]$ then it has all points in $J(f)$. If this periodic orbit is a stable fixed point x, and $x > 0$ then by 3, either $|f'|\big|_{[0,x]} \le 1$ or $|f'|\big|_{[x,1]} \le 1$. In the first case, x attracts the whole interval $[0,x]$ and hence 0, and in the second case it attracts all of $[x,1]$ and hence 1, and hence $0 = f^{-1}(1)$. Similarly if $x < 0$. If there is a stable periodic orbit with period $p \ge 2$ then a similar argument shows that the rightmost fixed point of f^p attracts either a critical point of f^p and hence 0 or it attracts 1. So there is no periodic orbit in $J(f)$ attracting only -1.

Consider now the case of a stable periodic orbit not attracting 0 or 1. It attracts -1, and hence there can

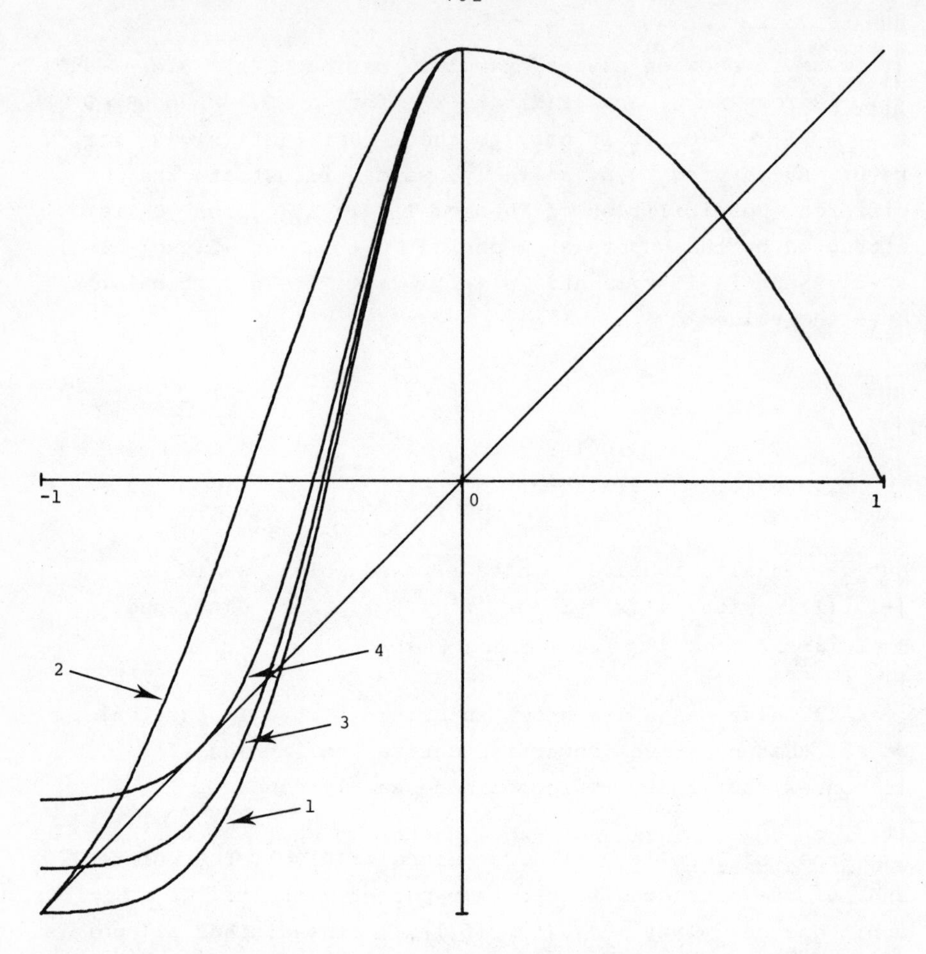

Figure II.8. $f(x)$ as described in text.

1: The point -1 is a stable fixed point, $f'(-1) < 1$.

2: The point -1 is a stable fixed point, $f'(-1) = 1$. -1 attracts only itself.

3: There is a fixed point which attracts -1 but not 0, 1, it attracts an open set.

4: There is a fixed point which attracts -1 but not 0,1. It attracts only from the left.

be at most one such. We claim it can be nothing else than a stable fixed point in $J(f)^{\perp} = [-1,f(1))$. If one point of the periodic orbit is in $J(f)$, then the whole period must be in $J(f)$ and it attracts 0 and 1 as we have seen. By 3, f can have at most two fixed points in $J(f)^{\perp}$, because it has at most two in $[-1,0]$. Assume it has none. Then $f(y) > y$ for $y \in J(f)^{\perp}$, and hence f can have neither a fixed point nor a periodic orbit in $J(f)^{\perp}$. Next assume f has exactly one fixed point x in $J(f)^{\perp}$. If $f'(x) > 1$ then we must have $x = -1$ because $f(y) < y$ for $y < x$. If $0 < f'(x) \leq 1$ it attracts -1 (since it cannot attract anything else). Hence f has a stable fixed point in $J(f)^{\perp}$ attracting -1, and hence no other periodic orbit in there. Finally assume f has two fixed points in $J(f)^{\perp}$. At one, we have $0 < f'(x) < 1$ and hence it attracts -1 as before. This proves Corollary 2.

Proof of Corollary II.4.4. If $f'(-1) > 1$, then $f(x) > x$ for all $x \in (-1,0)$. Hence there is no stable periodic orbit in $[-1,f(1))$.

We now wish to abandon some of the differentiability assumptions in the proof. Note first of all that if f is S-unimodal then $|f'|^{-1/2}$ is convex on $[-1,0)$ and on $(0,1]$. This can be seen by differentiating twice. Note that this is not equivalent to the concavity of $|f'|$. An even weaker condition, which is implied by $Sf < 0$ is the following.

Let $x_1 < x_2 < x_3 < x_4$ be points in a segment on which f is monotone. Denote by $R(x_1,x_2,x_3,x_4)$ the crossratio $(x_4-x_1)(x_3-x_2)/(x_4-x_3)(x_2-x_1)$. If $Sf < 0$, then $R(x_1,x_2,x_3,x_4) < R(f(x_1),f(x_2),f(x_3),f(x_4))$.

We shall say that f has property R if f is continuous, piecewise monotone and if the above inequality holds on every interval on which f is monotone. This definition has the advantage that it does not need differentiability. Nevertheless, most results for S-unimodal maps hold for $\mathscr{C}^1$-unimodal maps with property R. We list some of them now and indicate where the proofs differ from the S-unimodal case.

1'. If f and g have property R, then $f \circ g$ has property R. The proof is obvious from the definition.

2'. If f has property R then f^n has property R for all $n \geq 1$.

3'. If f is $\mathscr{C}^1$ (i.e., once continuously differentiable) satisfies condition R, then f' cannot equal 1 on an interval. This is obvious since f is then locally linear.

3". If f is $\mathscr{C}^1$, has finitely many critical points and has property R, and f' is piecewise monotone then $|f'|$ has no positive local minimum.

Proof. Assume $f' > 0$, and assume $x = 0$ is a positive local minimum of f', $f'(0) = a$. Consider a neighborhood of 0 in which f' is piecewise monotone. Then the graph of f' is about like this

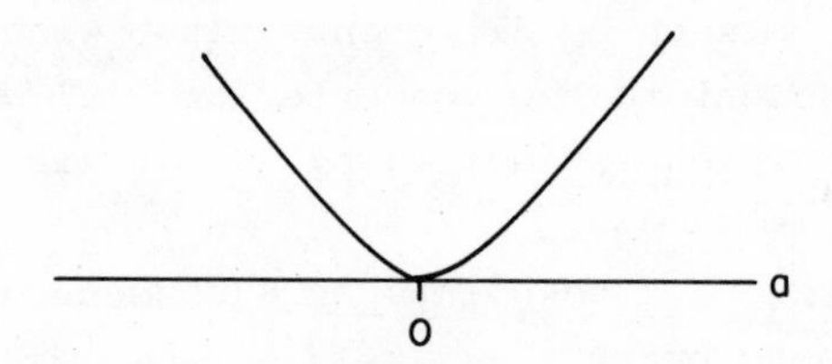

Figure II.9.

Thus in a neighborhood of 0, the graph of f looks like

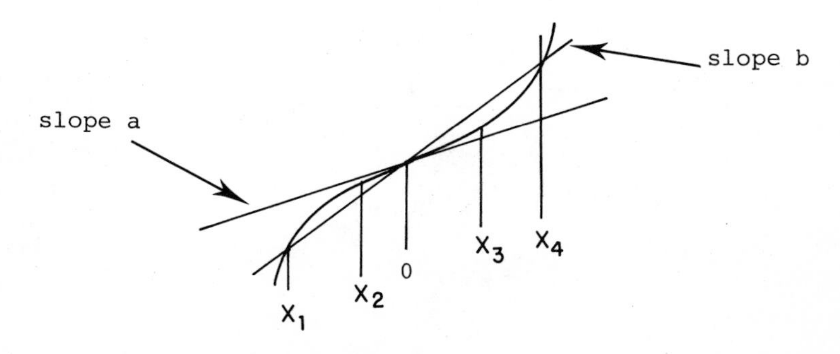

Figure II.10.

Choosing now x_1, x_2, x_3, x_4 as shown, we get for x_2, x_3 sufficiently near to zero

$$\frac{f(x_4)-f(x_1)}{x_4-x_1}\;\frac{f(x_3)-f(x_2)}{x_3-x_2} - \frac{f(x_4)-f(x_3)}{x_4-x_3}\;\frac{f(x_2)-f(x_1)}{x_2-x_1}$$

$$< b(a+\varepsilon) - (b+\varepsilon)^2 < 0 \quad .$$

When ε is sufficiently small, we arrive at a contradiction. The conclusions of 4, 5, 6, 7 hold now under the assumptions of 3". We have thus shown:

THEOREM II.4.5. *The conclusions of Theorem 1 and of Corollaries 2 and 4 hold for the class of $\mathscr{C}^1$-unimodal functions with piecewise monotone first derivative and which satisfy property* R *and* S4.

This will provide us later with the possibility to work with functions which are only piecewise $\mathscr{C}^2$. Another useful criterion to check whether $Sf < 0$ is the following

LEMMA II.4.6. *If* f *is a polynomial of degree* ≥ 2 *and all zeros of* f' *are real then* $Sf < 0$.

Proof. The assumption means $f'(x) = A\,\Pi_{j=1}^{n}(x-a_j)$, $a_j \in \mathbb{R}$. Then

$$Sf(x) = 2\sum_{i<j} \frac{1}{(x-a_i)(x-a_j)} - \frac{3}{2}\left[\sum_i \frac{1}{(x-a_i)}\right]^2$$

and this is manifestly <0 when the a_i are real. In particular all $f_\mu(x) = 1-\mu x^2$ satisfy $Sf_\mu < 0$ (since $f'''_\mu = 0$).

Remarks and Bibliography. As pointed out in this section, the question of the number of stable periodic orbits was addressed by Julia [1918]. To our knowledge, the role of the negative Schwarzian derivative for the question of the number of fixed points was discovered by D. Singer [1978]. For the relation between the Schwarzian derivative and analytic functions, see e.g. Hille [1976]. The role of the cross-ratio is pointed out in Guckenheimer [1979].

II.5 HOMTERVALS

So far, we have analyzed in some detail the relationship between itineraries and points in $[-1,1]$. We now ask the question of how many points can have the same itinerary. The answer to this question will be relevant for the sections on topological conjugacy, on sensitive behavior and on ergodic properties. It turns out that in the case of S-unimodal maps there are essentially two possibilities for the relationship between itineraries and points. If f is S-unimodal and has no stable periodic orbit, then we shall see that two different points always have different itineraries. On the other hand, if f is S-unimodal and has a stable periodic orbit, then almost all points (in the Lebesgue sense) have itineraries which are eventually periodic with the periodic part equal to the itinerary of a point of the periodic orbit (except in the superstable case).

The study of the above questions was simultaneously pushed through by Guckenheimer [1979] and Misiurewicz [1980]. Our presentation is a combination of the arguments of these papers. The most economic way to study the question of itineraries seems to be achieved by introducing the notion of homtervals. We define a homterval for f to be an open interval J such that $f^n|_J$ is a homeomorphism onto its image for all $n \geq 1$. Thus two points in a homterval have the same itineraries. On the other hand, by Lemma II.1.3, if $x < x'$ and x and x' have the same itinerary, then (x,x') is a homterval. Note that if J is a homterval, then $f^k(J)$ is a homterval for all $k > 0$. We say J is a homterval on V if J is a homterval and $f^j(J) \subset V$ for all $j \geq 0$.

Let W be an open subset of $(-1,0) \cup (0,1)$. We say $f|_W$ has a sink if there is an open interval $K \subset W$ such that $f^n(K) \subset K$ for some $n \geq 1$ and $f^j(K) \subset W$ for $j = 1,\ldots,n-1$ (and hence for all j). We note the following easy

LEMMA II.5.1. <u>If f is $\mathscr{C}^1$-unimodal and if W is an open subset of</u> $(-1,0) \cup (0,1)$ <u>such that $f|_W$ has a sink, then $\bar{W}$ contains a stable periodic orbit of</u> f.

<u>Proof</u>. If $f|_W$ has a sink K, then $f^n|_K$ is a homeomorphism of K into itself. Thus $g = f^{2n}|_K$ is an orientation preserving homeomorphism of K into itself. We want to show g has a stable fixed point in $\bar{K}$.

It is obvious that there is an $x \in \bar{K}$ such that $g(x) = x$. By the definition of K, $g'|_K \neq 0$. If g maps $\bar{K}$ into K then the smallest fixed point x in $\bar{K}$ of g lies in K and $0 < g'(x) \leq 1$, see Figure II.11: (The inequality $g'(x) > 0$ follows since g preserves orientation.)

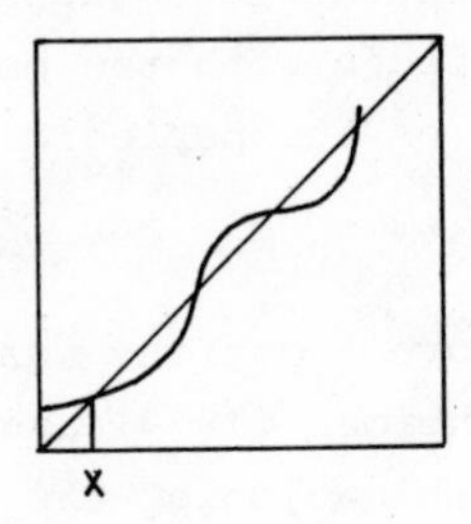

Figure II.11.

If $g(x) = x$ for some $x \in \partial\bar{K}$ then we argue as follows. Assume x is the left end of $\bar{K}$. If $g'(x) < 1$ we are finished. If $g'(x) = 1$ and $g(y) < y$ for y near x in K then the point x is (one-sided) stable. If $g(y) > y$ or $g'(x) > 1$ then we argue as in the case $g(\bar{K}) \subset K$. The upper end of $\bar{K}$ is handled in a symmetrical fashion.

Thus sinks adequately describe stable periodic orbits. We now restrict the geometrical set-up as follows: Let U be an open interval containing zero. Let U_1 be a closed interval containing zero, $\partial U_1 \not\ni 0$, and contained in U, and set $V = (-1,1) \setminus U_1$. When talking about U and V below, we shall always assume this geometrical relationship between the two sets.

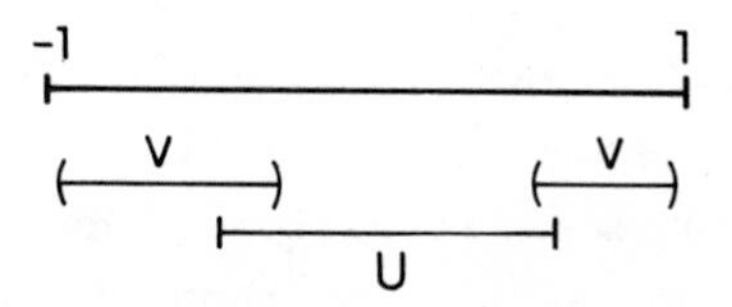

Figure II.12.

The set V will play the role of W described above, but our definitions now require $\mathrm{dist}(V,0) > 0$, while in Lemma 1 such a condition was not needed. In particular one can choose W (but not V) equal to $[-1,1] \setminus \{0,-1,1\}$.

THEOREM II.5.2. Misiurewicz [1980]. Let f be S-unimodal and define V as above. If $f|_V$ has no sink, then for any fixed open interval U with $U \cup V = [-1,1]$ the following are true.

1. For every homterval J on V there is an $m > 0$ such that $f^m(J) \subset U$.

2. There is an m such that for all x for which $f^j(x) \notin U$ for $j = 0,1,2,\ldots,m-1$, one has $|Df^m(x)| > 1$.

3. Define $E_n = \{x\colon f^j(x) \notin U$ for $j = 0,1,\ldots,n-1\}$. There is an $\eta < 1$ such that the Lebesgue measure of E_n is bounded by $\mathcal{O}(\eta^n)$.

This theorem is a detailed description of how a map behaves on orbits which avoid a neighborhood of the critical point. A special case of such orbits occurs for piecewise linear maps $g_\mu(x) = 1 - \mu|x|$, when the orbit of x avoids a neighborhood of 0. If $\mu > 1$ then g_μ has no sinks so that we can ask questions analogous to the conclusions 1, 2, 3, above. For example, the conclusion 2 is trivial with $m = 1$

in the case of piecewise linear maps, and a little reflection shows that (3) holds with $\eta = 1/\mu$.

The true content of the theorem is thus the observation that if one talks about points whose orbit $\{x, f(x), \ldots, f^n(x), \ldots\}$ avoids a neighborhood of the critical point, then if we "wait" for sufficiently many iterations, the conclusions for S-unimodal maps are very similar to those for piecewise linear maps. The nontrivial part of this observation comes from the fact that $|f'|_V|$ is not >1 in general, so that there are contracting regions for f in V. The following lemma will be useful for the proof of Theorem 2.

LEMMA II.5.3. *If f is $\mathscr{C}^1$-unimodal and J is a homterval then the intervals $f^n(J)$, $n = 0,1,2,\ldots$ are pairwise disjoint or there is a W such that $f|_W$ has a sink.*

Proof. Suppose that for some $n \geq 0$ and $k > 0$, $f^n(J)$ and $f^{n+k}(J)$ are not disjoint. Then for every $p \geq 0$, $f^{n+pk}(J)$ and $f^{n+(p+1)k}(J)$ are not disjoint. Thus, $K = \cup_{p\geq 0} f^{n+pk}(J)$ is an interval. For every p, the map

$$f^k|_{f^{n+pk}(J)} = (f^{n+(p+1)k}|_J) \circ (f^{n+pk}|_J)^{-1}$$

is a homeomorphism since J is a homterval. Thus $f^k|_K$ is a homeomorphism and $f^k(K) \subset K$, by construction. Then $f|_L$ has a sink, where $L = \cup_{j=0,\ldots k-1} f^j(K)$.

Proof of Theorem II.5.2. Since f is S-unimodal, and in particular $f'(-1) \neq 0$ and $f'(1) \neq 0$, we find from $\text{dist}(V,0) > 0$ that $\log|f'|$ is a Lipschitz function on every component of V. Let γ denote the Lipschitz constant of $\log|f'|$. We shall show first:

I. *If K is an interval such that $f^j(K) \subset V$ for $j = 0, \ldots, n-1$ then*

$$\log \frac{\sup_K |Df^n|}{\inf_K |Df^n|} \leq \gamma \sum_{k=0}^{n-1} \lambda(f^k(K)),$$

where λ *is Lebesgue measure.*

Proof. If $a, b \in K$ then we have

$$\log|Df^n(a)/Df^n(b)| = \sum_{k=0}^{n-1} \log|f'(f^k(a))| - \log|f'(f^k(b))|$$

Since $f^k(K)$ is an interval and is contained in V, it is contained in some component of V. Hence

$$\begin{aligned} \log|f'(f^k(a))| - \log|f'(f^k(b))| &\leq \gamma|f^k(a) - f^k(b)| \\ &\leq \gamma\lambda(f^k(K)) \end{aligned}$$

and I follows.

II. *If there is a homterval* J *on* V *such that* $\overline{f^n(J) \setminus U} \neq \emptyset$ *for all* $n \geq 0$, *then there is an* n_0 *such that for* $n \geq n_0$ *every homterval in* V *containing* $f^n(J)$ *is disjoint from* U.

Proof. Assume II is false. Then there is an $n \geq 0$ and a $k > 0$ and there are two homtervals K, L on V such that $f^n(J) \subset K$, $f^{n+k}(J) \subset L$ and both $\bar{K}$ and $\bar{L}$ contain the same endpoint of U and a piece of V adjacent to this endpoint. Thus $K \cup L$ is a homterval, and $f^k(K \cup L)$ intersects $K \cup L$. This contradicts Lemma 3. Hence II is true.

Assume there is a homterval J on V such that $\overline{f^n(J) \setminus U} \neq \emptyset$ for all $n \geq 0$ and let n_0 be defined by the conclusion of II. Let M be the maximal homterval on V containing $f^{n_0}(J)$. From II we have

III. <u>For each</u> $n \geq 0$, <u>every homterval on</u> V <u>containing</u> $f^n(M)$ <u>is disjoint from</u> U. Denote $\beta = \mathrm{dist}([-1,1] \setminus U, [-1,1] \setminus V)$. By construction $\beta > 0$. We shall prove the following, after completing the proof of Theorem 2.1.

IV. <u>If</u> $L \subset V$ <u>is an open interval containing</u> M, $L \neq M$ <u>and such that</u> $\lambda(L)/\lambda(M) < (\beta/2)e^{-\gamma(\beta+2)} + 1$, <u>then</u>

$$\frac{\lambda(f^k(L))}{\lambda(f^k(M))} < \frac{\beta}{2} + 1$$

<u>and</u> $f^k|_L$ <u>is a homeomorphism for all</u> $k \geq 0$.

From $(\beta/2)e^{-\gamma(\beta+2)} + 1 > 1$ and III we deduce that there is a $L \neq M$ satisfying the hypotheses of IV. This L is, by IV, a homterval on V and this contradicts the maximality of M. Thus Theorem 2.1 is proven. It remains to show IV. We proceed by induction on k. The case $k = 0$ is obvious. Suppose IV has been shown for $k = 0, 1, \ldots, n-1$. We have then $\lambda(f^k(L) \setminus f^k(M)) < \beta\lambda(f^k(M))/2 \leq \beta/2$. From this, from III and the definition of β it follows that $f^k(L) \subset V$ for $k = 0, \ldots, n-1$. Hence, since L is an interval, $f^n|_L$ is a homeomorphism. By I, and Lemma 3, we obtain

$$\frac{\lambda(f^n(L))}{\lambda(f^n(M))} - 1 = \frac{\lambda(f^n(L \setminus M))}{\lambda(f^n(M))} < \frac{\lambda(L \setminus M)}{\lambda(M)} \cdot e^{\gamma \sum_{k=0}^{n-1} \lambda(f^k(L))}$$

$$\leq \frac{\beta}{2} e^{-\gamma(\beta+2)} e^{\gamma \sum_{k=0}^{n-1} \left(\frac{\beta}{2} + 1\right) \lambda(f^k(M)}$$

$$\leq \frac{\beta}{2} e^{\gamma\left[\left(\frac{\beta}{2} + 1\right) 2 - \beta - 2\right]} = \frac{\beta}{2} .$$

<u>Proof of Theorem 2.2</u>. Suppose that for every $n \geq 1$ there exists x_n such that $|Df^n(x_n)| \leq 1$, and $f^j(x_n) \not\in U$

for $j = 0,1,\ldots,n-1$. We shall derive a contradiction from this. Define J_n as the maximal open subinterval containing x_n, and such that $f^j(J_n) \subset V$ for $j = 0,\ldots,n-1$. The point x_n divides J_n into two subintervals. Since f is S-unimodal, we have by (3) in the proof of Theorem II.4.1, that $|Df^n| \le 1$ on one of the two subintervals. Denote this subinterval by L_n. By the maximality of J_n, there is a $k(n) < n$ such that

I. $f^{k(n)}(L_n)$ <u>has a common endpoint with a component of</u> V.
We claim

II. $\lambda(L_n) \to 0$ <u>as</u> $n \to \infty$.

<u>Proof of II</u>. Assume the contrary. Then we can choose a sequence $n_1 < n_2 < \ldots$, and an $\varepsilon > 0$ such that

(i) The x_{n_i} converge to a point x_0, when $i \to \infty$.

(ii) $\lambda(L_{n_i}) > \varepsilon$.

(iii) All L_{n_i} are on the same side of x_{n_i}.

(iv) $|x_{n_i} - x_0| < \varepsilon/2$ for all i.

For every j there is an $i(j)$ such that $n_{i(j)} > j$. From the continuity of f it follows that $f^j(x_{n_i}) \not\in U$ for $i \ge i(j)$ implies $f^j(x_0) \not\in U$ for all j. In fact, $f^j(x_0)$ is on the same side of 0 as $f^j(x_{n_i})$ for i sufficiently large. Consider now the open interval X whose one endpoint is x_0 and whose other endpoint lies in (all) L_{n_i}, and which has length $\varepsilon/2$. Such an interval exists by (iii) and (iv). From $f^j(L_{n_{i(j)}}) \subset V$ and $f^j(x_0) \not\in U$ it follows that X is a homterval on V. But $f^j(\overline{X}) \not\subset U$ for all j and this contradicts Theorem 2.1. Thus II is true.

From $|Df^n|_{L_n}| \le 1$ we have $\lambda(f^n(L_n)) \le \lambda(L_n)$ and hence from II,

$$\lambda(f^{n-k(n)}(f^{k(n)}(L_n))) \to 0 \qquad \text{as} \quad n \to \infty . \qquad (*)$$

From I, and from $f^{k(n)}(x_n) \notin U$, we see that $f^{k(n)}(L_n)$ contains one of the two components of $U \cap V$ (unless $k(n) \leq 1$, a trivial variant). Thus there is one of these components, K, and a sequence $n_1 < n_2 < \ldots$, such that $f^{k(n_i)}(L_{n_i}) \supset K$ for all i. Thus (*) implies $\lambda(f^{n_i - k(n_i)}(K)) \to 0$ as $i \to \infty$, and thus $n_i - k(n_i) \to \infty$ as $i \to \infty$. Let a be one of the accumulation points of the set $\{f^{k(n_i)}(x_{n_i})\}_{i=1}^{\infty}$. Since $f^j(f^{k(n_i)}(x_{n_i})) \notin U$ for $n_i - k(n_i) > j$, we obtain from $n_i - k(n_i) \to \infty$ that $f^j(a) \notin U$ for all j. Let M be the minimal open interval such that $K \cup \{a\} \subset \bar{M}$. Then M is a homterval on V. But by Theorem 2.1 this contradicts $f^j(a) \notin U$ for all j. This proves Theorem 2.2.

<u>Proof of Theorem 2.3</u>. By Theorem 2.2 there is an m such that for all $y \in V$ with $f^j(y) \notin U$ for $j < m$ we have $|Df^m(y)| > 1$. By making V slightly smaller such that $V \cap U$ still has the required form, we see that $|Df^m(y)| \geq \alpha > 1$. Let $\rho \leq \inf_{x \notin U} |f'(x)|$, we can choose $1 > \rho > 0$. Let γ be the Lipschitz constant for $\log|f'|\,|_V$.

For $n = pm + q$, with $0 \leq q < m$, we have for $x \in E_n$:

$$|Df^n(x)| = \left| \prod_{j=0}^{p-1} Df^m(f^{jm}(x)) \prod_{\ell=0}^{q-1} f'(f^{\ell+pm}(x)) \right| \geq \alpha^p \rho^q .$$

Let $\beta = \alpha^{1/m} > 1$. Then $|Df^n(x)| \geq \beta^{mp}\rho^q \geq \alpha^{-1}\beta^n\rho^m$ if $x \in E_n$. But then for any $k \leq n$, and $x \in f^k(E_n) \subset E_{n-k}$,

$$|Df^{n-k}(x)| \geq \frac{1}{\alpha}\beta^{n-k}\rho^m .$$

Therefore, for every component K of E_n

$$2 \geq \lambda(f^n(K)) \geq \frac{1}{\alpha}\beta^{n-k}\rho^m \lambda(f^k(K)) .$$

Thus

$$\lambda(f^k(K)) \le 2\alpha\beta^{k-n_\rho-m} .$$

We can now bound, as on page 111,

$$\log\left(\sup_{x\in K} |Df^n(x)| \,/\, \inf_{x\in K} |Df^n(x)|\right) \le \gamma \sum_{k=0}^{n-1} \lambda(f^k(K))$$

$$\le 2\alpha\rho^{-m} \sum_{i=1}^{n} \left(\frac{1}{\beta}\right)^i \le 2\alpha\rho^{-m}/(\beta-1) = \log\delta \qquad (**)$$

To simplify the notation, we consider only the case when f, U and V are symmetric. Let $U = (a,-a)$ and let $r > 0$ be the smallest integer for which $f^r(a) \in U$. We first pursue the case $r < \infty$. Define $\mu = \min\{|a-f^r(a)|, |-a-f^r(a)|\}$. Consider a connected component K of E_n. We do not consider here the trivial variant when $K \ni \pm 1$. In all other cases, $K = (u,v)$ and there are maximal $j,k < n$ such that $f^j(u) \in \partial U$ and $f^k(v) \in \partial U$. Since r is finite, there is a p, $n \le p \le r+n-1$ for which $f^p(K)$ first hits U. It follows that $\lambda(f^p(K \setminus E_{p+1})) \ge \mu$. Thus $\lambda(K \setminus E_{p+1}) \ge \mu(\sup_{x\in K} |Df^p(x)|)^{-1}$. From $E_{r+n} \subset E_{p+1}$ we have $\lambda(K \setminus E_{r+n}) \ge \mu(\sup_{x\in K} |Df^p(x)|)^{-1}$. Since $f^p|_K$ is a homeomorphism and $f^p(K) \subset [-1,1]$ we also have $\lambda(K) \le 2\,(\inf_{x\in K} |Df^p(x)|)^{-1}$. Thus

$$\frac{\lambda(K \cap E_{r+n})}{\lambda(K)} \le 1 - \frac{\lambda(K \setminus E_{r+n})}{\lambda(K)} \le 1 - \frac{\mu \inf_{x\in K}|Df^p(x)|}{2\sup_{x\in K}|Df^p(x)|}$$

$$< 1 - \mu/(2\cdot\delta) \equiv \eta^r .$$

Summing over K, the assertion follows in this case.

Next we consider the case $r = \infty$, and we want to reduce it to the preceding one by choosing a smaller interval U_1 which still intersects V. If $f(U)$ is a homterval, we have by Theorem 5.2.1 that $f^m(U) \subset U$ for some $m < \infty$, which contradicts $r=\infty$. Hence $f(U)$ is not a homterval and there is a smallest $s>1$ for which $f^s(U)\cap U\neq\emptyset$, and in fact $f^s(U)\supset U$. Assume for definiteness that $f^s(a)<a<0$, and let L be a closed interval contained in $U\cap V$ with $a\in\partial L$ and $\mathrm{dist}(\partial L,\partial V)>0$. If $|f^i(x)|>|x|$ for all $x\in L$ and all $i>0$ then $f^i(L)\subset V$ for all i. This means $f(L)$ is a homterval on V. Applying Theorem 5.2.1 with an interval V and U_2 as shown in the figure below, we see that some iterate of $f(L)$

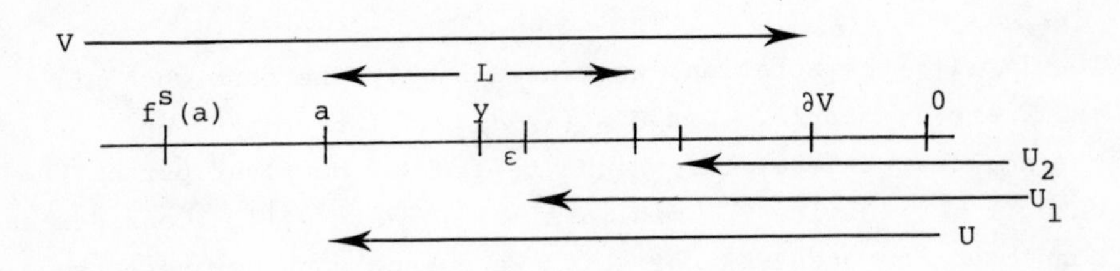

must intersect U_2, a contradiction. Hence there is an i and an x for which $|f^i(x)|<|x|$. Choosing the smallest such i and since $f^i(a)\notin U$ we see that $f^i(y)=y$ for a $y\in L$. This y is an unstable fixed point. We choose $U_1=(y+\varepsilon,-y-\varepsilon)$ with $\varepsilon>0$ small, so that $y+\varepsilon\in V$. Making ε smaller, if necessary, we can achieve $f^{2i}(\partial U_1)\subset U_1$, since y is repelling. Hence we are reduced to the case of finite r as asserted. This completes the proof of Theorem 5.2.3.

We are going to apply now the general Theorem 2 to some specific situations, which shall be relevant later.

THEOREM II.5.4. _Let_ f _be_ S-_unimodal and suppose_ f _has no stable periodic orbits_. _Then_ f _has no homterval_.

This implies that different points have different itineraries. Since the boundaries of homtervals are contained in the closure of the set of _pre-images_ of the critical point 0 (i.e., by points y such that $f^k(y)=0$ for some $k\geq 0$) we have the following

COROLLARY II.5.5. _If_ f _is_ S-_unimodal and has no stable periodic orbit, then the set of pre-images of_ 0 _is dense in_ $[-1,1]$.

Proof of Theorem II.5.4. Assume f has a homterval J, but no stable periodic orbit. We shall derive a contradiction from this. By Lemma 3, at most two among the sets $f^m(J)$, $m \geq 0$ can have 0 as an endpoint. If k is the largest m for which this occurs, then we consider $f^{k+1}(J)$ (which is again a homterval) instead of J. Thus we may assume without loss of generality that if f has a homterval, then it has another homterval J such that none of the $f^m(J)$, $m \geq 0$ has 0 as an endpoint.

By Lemma 1, f satisfies the assumptions of Theorem 2 for every V (of the standard form) with $\operatorname{dist}(V,0) > 0$. Thus if K is a homterval, then for every open interval U containing 0 there is, as a consequence of Theorem 2.1, an m such that $f^m(K) \subset U$.

We pursue the argument for the special case when f is a symmetric function and refer the reader to the original reference for the general case. By the preceding argument, choosing U sufficiently small, we see that there is for every homterval K a smallest integer $m = m(K)$ such that $f^m(K)$ is closer to zero than K, that is $\operatorname{dist}(K,0) > \operatorname{dist}(f^m(K),0)$. Recall that Lemma 3 implies $K \cap f^m(K) = \emptyset$. We shall now argue that $\lambda(f^{m(K)}(K)) \geq \lambda(K)$. Since we can iterate the argument indefinitely, starting from the first homterval J, and since $f^m(J) \not\ni 0$ by construction, we find that the combined volume of all homtervals must exceed any upper bound, and this leads to a contradiction. The key to proving $\lambda(f^{m(K)}(K)) \geq \lambda(K)$ is the following lemma.

We shall also introduce some new notation. Let $x \in (-1,1)$. Then x' will be the other point in $[-1,1]$ for which $f(x') = f(x)$. There can be at most one such point.

If there is such a point, then $(x;x')$ will denote the open interval whose endpoints are x and x'. If there is no such point and $x \neq 0$, then we set x' equal to $+1$ or -1 and $(x;x')$ coincides with $\{y \mid f(y) > f(x)\}$. We shall say $f(x)$ is closer to 0 than x if $f(x) \subset (x;x')$. We extend the notation $(x;y)$ to denote the open interval whose endpoints are x and y, and by analogy to half-open and closed intervals. Let us use the notation $p^{\#}$ to denote p or p'.

LEMMA II.5.6. Let f be unimodal and define $K_n = \{x\colon f^i(x) \notin [x;x']$ for $i = 1,2,\ldots,n-1$ and $f^n(x) \in (x;x')\}$. Every connected component of K_n is of the form (p,q) where $f^n(p) = p^{\#}$ and $f^n(q) = q^{\#}$. If f is S-unimodal and has no stable periodic orbit then $f^n(p) = p$ and $f^n(q) = q'$ (or $f^n(p) = p'$ and $f^n(q) = q$) and f^n is monotone on every component of K_n.

Proof. Consider a connected component of K_n. We concentrate on one of its endpoints, p. There is a smallest i, $1 \leq i \leq n$ for which $f^i(p) = p^{\#}$. We claim this implies $f^n(p) = p^{\#}$. Since $f(p') = f(p)$, we see that the claim is true if $n = ki$ for some integer k. Otherwise, $n = ki + j$ with $1 \leq j < i$. Then we find $f^n(p) = f^j(p^{\#}) = f^j(p)$. But this is impossible, since $f^j(p) \notin [p;p']$ and $f^n(p) \in [p;p']$.

Next we consider the stronger assumptions of S-unimodal f without stable periodic orbit. From $f^n(0) \neq 0$ we have $0 \notin \overline{K_n}$. For $x \in K_n$ we have $f^i(x) \neq 0$ if $x \neq 0$ and $i < n$, and thus we find that f^n is monotone on components of K_n. Furthermore, K_n does not have critical points of f^n as endpoints of its components. Thus f^n is strictly monotone on components of K_n. Since f^n has no stable fixed point, it follows from $f^n(p) = p^{\#}$ and $f^n(q) = q^{\#}$ that $p^{\#} = p$ and $q^{\#} = q'$ or $p^{\#} = p'$ and $q^{\#} = q$, cf. Fig. II.13.

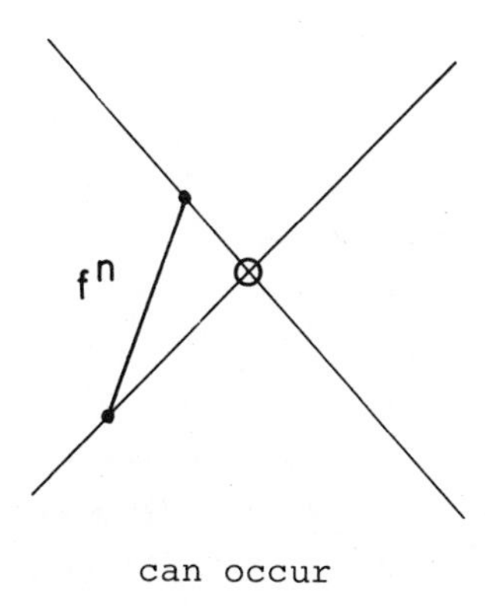

can occur

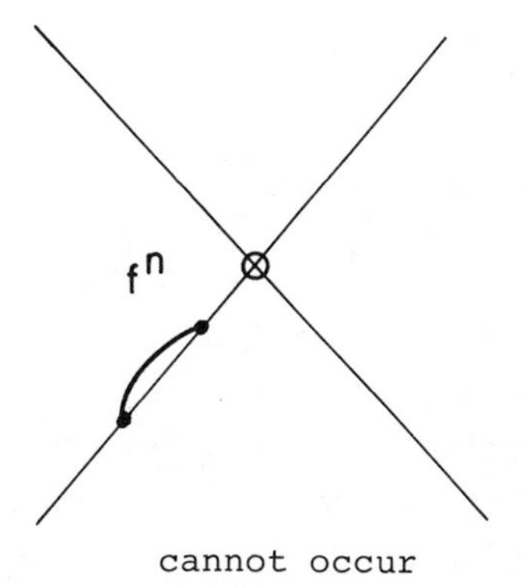

cannot occur

Figure II.13.

We now complete the proof of Theorem 4. If J is a homterval, and $m(J)$ is the first iterate for which $f^{m(J)}(J)$ is closer to zero than J, then we see that $J \subset K_{m(J)}$ where K_m is defined as in Lemma 6. (Use Lemma 3.) Denote by S the connected component of $K_{m(J)}$ in which J lies. Then $S = (p,q)$ with $f^m(p) = p$ and $f^m(q) = q'$, $m = m(J)$ (or the primes reversed). Assume the first alternative. Then p is a fixed point of f^m and since f has no stable periodic orbit, we have $|Df^m(p)| > 1$. But we also have, when f is symmetric, $|Df^m(q)| = |Df^m(q')|$ and $|Df^m(q')| > 1$ since q' is also a periodic point of f. (If f is not symmetric, one needs here $f''(0) \neq 0$ or some similar condition to bound the ratio of $f^m(q)$ and $f^m(q')$, cf. Guckenheimer [1979].) By Lemma 6, f^m is monotone on S and hence $|Df^m| > 1$ on all of S, since $|Df^m|$ does not have a local minimum. Hence $\lambda(f^{m(J)}(J)) \geq \lambda(J)$. This completes the proof of Theorem 4.

We next discuss the case when f has a stable periodic orbit. We define $E_f = \{x\colon f^n(x)$ does not tend to the stable periodic orbit of $f\}$. The next proposition shows that E_f is a set of exceptional points.

PROPOSITION II.5.7. _If_ f _is_ S-_unimodal and has a stable periodic orbit, then_ $\lambda(E_f) = 0$.

This implies that E_f contains no non-trivial intervals.

Proof. Define the stable neighborhood U of 0 as the maximal interval containing 0 (i.e., the critical point of f) and consisting of points which tend to the periodic orbit. There are two possibilities, according to the discussion in Section 4. The first case occurs when the stable periodic orbit is stable from both sides, cf Fig. II.14a) below, while the second occurs if it is stable from one side only (cf. Fig. II.14.b).

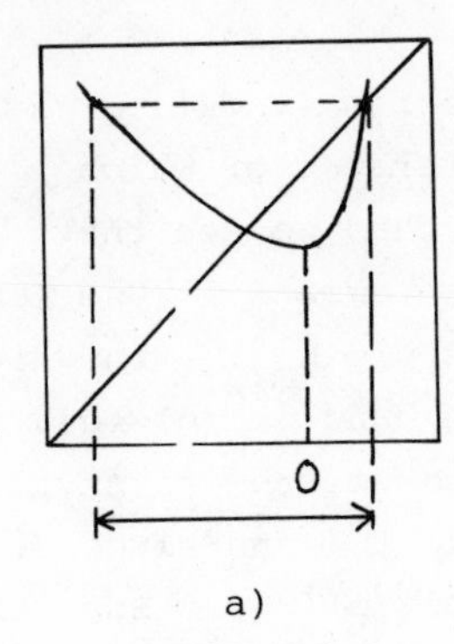

a)

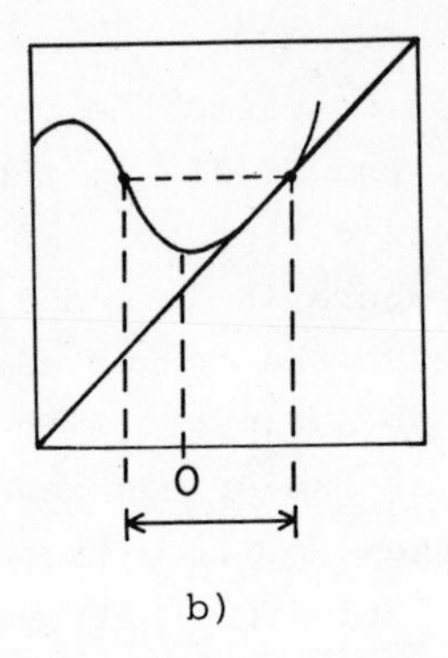

b)

Figure II.14. Graph of f^n (locally), when the stable periodic orbit has period n.

We see that U is open in the two-sided case. Thus U is of the form $(a;a')$. Note that $E_f = \cap_{n\geq 0} E_n(U)$, with $E_n(U)$ as defined in Theorem 2.3. By definition, $f|_V$ has no sink where $V = (-1,1)\setminus U_1$ with $0 \in U_1$ a closed interval in U. Then Theorem 2.3 implies $\lambda(E_{f|_V}) = 0$ which implies the result. The proof in the one-sided case needs a variant of Theorem 2.3, where $V = [-1,1]\setminus U$ and U is closed. We do not present this proof. Cf. Guckenheimer [1979].

COROLLARY II.5.8. If f is S-unimodal and has a stable periodic orbit, then two different points in E_f have different itineraries.

Proof. Assume the contrary, i.e., $\underline{I}(x) = \underline{I}(x')$, for some $x < x'$, $x,x' \in E_f$. Since all orbits of points in E_f avoid the set U considered in the proof of Proposition 7, (x,x') is a homterval, and thus for all $y \in (x,x')$, $f^n(y) \notin U$ for all n, because f is unimodal. This contradicts $\lambda(E_f) = 0$.

Remarks and Bibliography. The role of homtervals is already mentioned implicitly in Milnor-Thurston [1977]. Our presentation here is based to a large extent on the work of Misiurewicz [1980]. Lemma 6 and Theorem 4 are due to Guckenheimer [1979]. The proofs of Misiurewicz are slightly more elegant than Guckenheimer's, but also slightly weaker statements are being shown in his paper. Our choice is thus dictated by the desire of showing the common trend of these two papers, by working out the simpler proof and those parts of it which generalize readily to the more complicated situation described by Guckenheimer. This implies that some special cases are not worked out here and must be looked up in Guckenheimer [1979].

II.6 TOPOLOGICAL CONJUGACY

Two maps $f, g: [-1,1] \to [-1,1]$ will be called _topologically conjugate_ if there is a homeomorphism $h: [-1,1] \to [-1,1]$ which _preserves orientation_ such that

$$f = h^{-1} \circ g \circ h \quad .$$

We shall write $f \sim g$ if f and g are topologically conjugate. Obviously, $\sim$ is an equivalence relation. Let now f be unimodal and $f \sim g$; i.e., $f = h^{-1} \circ g \circ h$. Then g has all the properties of a unimodal map except that g takes its maximum at $h(0)$ whereas f takes its maximum at 0. We nevertheless want to associate an itinerary to g. So we extend the definitions of Section 1, in the obvious way, reserving the symbol C for $x = h(0)$, L for $x < h(0)$ and R for $x > h(0)$. Then we can define itineraries as before. All the results of the preceding sections carry over to continuous functions g whose maximum lies at $-1 < x_0 < +1$, $g(x_0) = 1$ and which are strictly monotonically increasing (resp. decreasing) on $[-1, x_0]$ (resp. $[x_0, 1]$). To simplify notation, we shall now again restrict attention to the case $x_0 = 0$. Now if $f \sim g$, with $h \circ f = g \circ h$, then $\underline{I}_f(x) = \underline{I}_g(h(x))$. In particular $\underline{I}_f(1) = \underline{I}_g(1)$. Thus we see that _topologically conjugate unimodal maps have the same kneading sequences_.

The aim of this section is to analyze the extent to which the converse is true. Does a kneading sequence already determine a map? This question was primarily addressed by Milnor-Thurston [1977], and a very satisfactory answer was given by Guckenheimer [1978]. This section is mainly a presentation of his results, combined with the analysis of Misiurewicz, which we presented in Section 5.

It is easy to see that if f, g are unimodal and $\underline{I}_f(1) = \underline{I}_g(1)$, this does not necessarily imply $f \sim g$. The following figure illustrates this: $\underline{I}_f(1) = \underline{I}_g(1) = RC$.

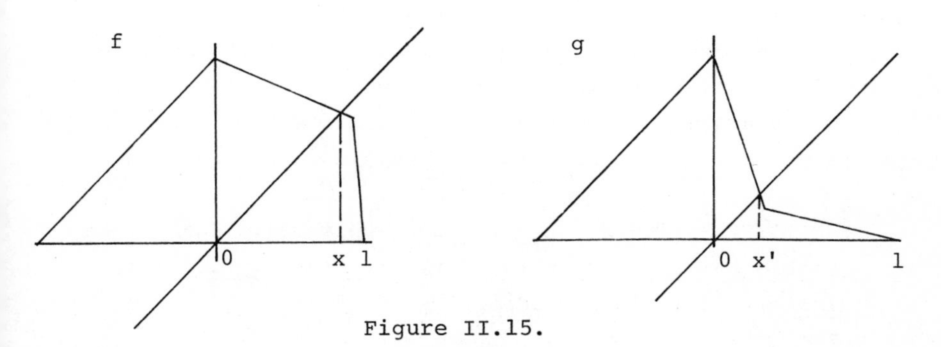

Figure II.15.

But these two functions cannot be topologically conjugate since the first has a stable fixed point, and the second has not. The example is revealing in two ways. First the itinerary of 1 does not "feel" the presence or absence of a stable periodic orbit (in contrast with the case of negative Schwarzian derivative). Second, the definition of itineraries does not allow one to measure the "size" of what is eventually attracted by the stable periodic orbit. Note that for <u>all</u> x, $\underline{I}_f(x) = \underline{I}_g(x)$ in the above example; the only choices are LC, LR^∞, C, R^∞, RC. (It may also be instructive to look at Theorem II.3.8 .)

It therefore looks most reasonable to prove topological equivalence when f, g have negative Schwarzian derivative and no stable periodic orbit. This is what we shall do now.

We shall always consider S-unimodal functions.

The first main result of this section is

THEOREM II.6.1. <u>Suppose that</u> f, g <u>are</u> S-<u>unimodal and</u> f <u>has no stable periodic orbit</u>. <u>If</u> $\underline{I}_f(1) = \underline{I}_g(1)$ <u>and</u>

$\underline{I}_f(-1) = \underline{I}_g(-1)$ *then* f *and* g *are topologically conjugate*.

We shall see in Proposition 2, that one can decide on the basis of $\underline{I}(1)$ alone whether or not a function has a stable periodic orbit in $[f(1),1] = J(f)$. The strategy of the proof is the following. We define a map $h: [-1,1] \to [-1,1]$ by the requirement $\underline{I}_f(x) = \underline{I}_g(h(x))$. This map is a homeomorphism if there is at most one point for f and for g assuming a given itinerary. (This will be the content of the analysis given below.) It is then clear that $h \circ f = g \circ h$.

PROPOSITION II.6.2. *Let* f *be* S-*unimodal*. *Then* f *has a stable periodic orbit in* J(f) *if and only if* $\underline{I}(1)$ *is finite or* $\underline{I}(1)$ *is periodic*.

Recall that two admissible sequences $\underline{A},\underline{B}$ are consecutive if there is nothing "between" $\underline{A}$ and $\underline{B}$, i.e., if $\underline{A} < \underline{B}$, then $\underline{A} \le \underline{X} < \underline{B}$ implies $\underline{X} = \underline{A}$ and similarly when $\underline{A} > \underline{B}$.

Proof of Proposition 2. Assume f has a stable periodic orbit of period n in J(f). Let x be a point of the stable periodic orbit. We can assume 0 is not on this periodic orbit since otherwise the assertion is obvious. Let y be a critical point of f^n such that $0 < |Df^n| \le 1$ on $(x;y)$. Let j be such that $f^j(y) = 0$. $f^{rn}(y)$ converges monotonously to x when $r \to \infty$. Since there is no critical point of f^n on $(x;y)$, $f^j|_{[x;y]}$ is a homeomorphism. Therefore $f^j(f^{rn}(y)) = f^{rn}(0)$ converges monotonously to $f^j(x)$. Assume now $0 \in (f^k(0); f^{k+j}(x))$ for some k. This implies $0 \in (f^{k+j}(y); f^{k+j}(x))$; in particular when $rn > k+j$ there is a $z \in (y;x)$ such that $Df^{rn}(z) = 0$. From $Df^{rn}(z) = \Pi_{t=0}^{r-1} Df^n(f^{tn}(z))$, $f^{tn}((x;y)) \subset (x;y)$ and $Df^n \neq 0$ on $(x;y)$, we deduce a contradiction. Hence $1 = f(0)$ and $f^{j+1}(x)$ have the same itinerary. Hence $\underline{I}(1)$ is periodic (cf. Lemma II.3.1) (and not only eventually periodic). If $\underline{I}(1)$ is finite, f has a superstable period. If $\underline{I}(1)$ is periodic of period p, then f^k is monotone on the interval

$[f^{p+1}(0),1]$ for all $k > 0$ since the endpoints of the interval have the same itineraries. Either $f^{p+1}(0) = 1$ in which case $f^p(0) = 0$ and $\underline{I}(1)$ was finite and we have a superstable orbit or $f^{p+1}(0) < 1$, and then $(f^{p+1}(0),1)$ is a homterval, and f has a stable periodic orbit by Theorem II.5.4. This proves Proposition 2.

Proof of Theorem II.6.1. Proposition 2 and the hypotheses of Theorem 1 imply that g has no stable periodic orbit. Let $x \in [-1,1]$. By Lemma II.1.3, we have $\underline{I}_f(-1) \le \underline{I}_f(x) \le \underline{I}_f(1)$ and if $x \neq x'$ then $\underline{I}_f(x) \neq \underline{I}_f(x')$ by Theorem II.5.4. From our assumptions, we find $\underline{I}_g(-1) \le \underline{I}_f(x) \le \underline{I}_g(1)$ and thus Theorem II.3.8 implies that there is a $y = h(x)$ such that $\underline{I}_g(y) = \underline{I}_f(x)$. Theorem II.5.4 implies y is unique. By Lemma II.1.2 and 3 and Theorem II.5.4, h is strictly monotone. Since h is 1-1 and onto, it is a homeomorphism. We list here a variant of Theorem 1.

THEOREM II.6.1.A. *Suppose* f,g *are* S-*unimodal and* f *has no stable periodic orbit. If* $\underline{I}_f(1) = \underline{I}_g(1)$ *then* $f|_{[f(1),1]}$ *and* $g|_{[g(1),1]}$ *are topologically conjugate through a homeomorphism* $\hat{h}$: $[f(1),1]$ *onto* $[g(1),1]$.

Proof. In the proof of Theorem 1, replace $\underline{I}_f(-1)$ by $\underline{I}_f(f(1))$ and $\underline{I}_g(-1)$ by $\underline{I}_g(g(1))$. Since $\underline{I}_f(1) = \underline{I}_g(1)$ implies $\underline{I}_f(f(1)) = \underline{I}_g(g(1))$ the proof goes as before.

In a way, Theorem 1A is more natural than Theorem 1 since f maps $[f(1),1]$ *onto* $[f(1),1]$. In the discussion of the conjugacy for stable periodic orbits, we shall thus restrict ourselves to $f|_{J(f)}$ with $J(f) = [f(1),1]$.

Another possibility, which we do not pursue here, would have been to consider f with $f(0) \le 1$, $f(1) = f(-1) = 0$. This case has been analyzed by Guckenheimer [1979]. What gets lost then is the nice relation between $\underline{I}_E(1)$ and $\underline{I}_E(0)$, which we have with our definitions. Of course, all this is just a matter of convenience.

We now state the summary of the topological classification of S-unimodal maps, as proved by Guckenheimer [1979]. The theorem shows that $\underline{I}_f(1)$ distinguishes topological equivalence classes except for one case in which a binary choice, not expressible through the itinerary $\underline{I}_f(1)$, has to be made.

THEOREM II.6.3. *Let* f,g *be S-unimodal, and assume* $\underline{I}_f(1) = \underline{I}_g(1)$.

1. *If* $\underline{I}_f(1)$ *is finite, then* $f|_{J(f)}$ *and* $g|_{J(g)}$ *are topologically conjugate.*

2. *If* $\underline{I}_f(1)$ *is infinite and periodic of period* n, $\underline{I}_f(1) = \underline{D}^\infty$, $|\underline{D}| = n$, *then there are two possibilities:*

   a. *If* $\underline{D}$ *is odd then* $f|_{J(f)}$ *and* $g|_{J(g)}$ *are topologically conjugate if and only if their stable periodic orbits have the same period* (which is n or $2n$).

   b. *If* $\underline{D}$ *is even, then* $f|_{J(f)}$ *and* $g|_{J(g)}$ *are topologically conjugate if their stable periodic orbits* (which have period n) *are both stable from one side or stable from both sides.*

3. *If* $\underline{I}_f(1)$ *is infinite but not periodic, then* $f|_{J(f)}$ *and* $g|_{J(g)}$ *are topologically conjugate.*

REMARKS. By Proposition 2, f and g have stable periodic orbits in Cases 1 and 2, and no stable periodic orbit in Case 3. Case 3 is a rewording of Theorem 1A. One can illustrate the various alternatives of Theorem 3 by the following examples.

We consider the maps f_μ^3, where $f_\mu(x) = 1 - \mu x^2$ with μ near 1.75. The global picture of the graph of f_μ^3 is presented in Fig. II.16 for three values of μ. Next we enlarge the region around $x=0$, $f_\mu^3(x)=0$, in Fig. II.17 for eleven values of μ. These values are equally spaced and are chosen in such a way that the curve No. 3 is tangent to the diagonal $f(x)=x$ and the curve No. 5 is superstable, i.e., $f_{\mu_5}^3(0)=0$. When the number of the curve increases, the derivative at the fixed point is seen to have a tendency to decrease and for the curve No. 11 we have in fact $Df_{\mu_{11}}^3(x_0)<-1$ where x_0 is defined by $f_{\mu_{11}}^3(x_0)=x_0$. This is maybe not very strongly visible in Fig. II.17. But in any case, it is now interesting to analyze f_μ^6. The global picture is shown in Fig. II.18, for the value of $\mu=1.7548\ldots$, i.e., the case when f_μ has a superstable period 3. In Fig. II.19, this is again enlarged for the same eleven values of μ as in Fig. II.17. Finally in Fig. II.20, we enlarge the left portion for $\mu_9, \mu_{10}, \mu_{11}$. It is clearly seen how the stable fixed point of f_μ^6, (which is actually the stable periodic orbit of period 3) loses its stability when passing from μ_{10} to μ_{11}. This is called a <u>bifurcation</u>. At μ_{11}, we find then two points of the stable period 6, with a unstable period 3 in between.

<u>Proof of Theorem 3</u>. By Theorem II.3.8, we know that the sets of itineraries for $f|_{J(f)}$ and $g|_{J(g)}$ are the same. Assume the periodic orbits of f and g are stable. Then by Corollary II.5.7, if $x\in E_f$, then there are no other points in $J(f)$ with the itinerary of x.

If U_f is the stable neighborhood of 0 for f, then $f^n(x)\in U_f$ iff $\mathscr{S}^{n+1}(\underline{I}_E(x))=\underline{I}_E(1)$. So we can determine whether $x\in E_f$ from its itinerary. The same considerations apply to g and hence we can define a homeomorphism $h:E_f\to E_g$ by the property that the $\underline{I}_f(x)=\underline{I}_g(h(x))$. We extend h to the set $\cup_{i>0} f^{-i}(U_f)$, assuming that we already have constructed a topological equivalence h from $f^n|_{U_f}$ to $g^n|_{U_g}$, where n is the smallest integer for which $f^n(U_f)\subset U_f$ (and hence also $g^n(U_g)\subset U_g$). For each component K of $f^{-i}(U_f)$ other than U_f there is a $j\le i$ with f^j mapping K homeomorphically into U_f. Then we define h for $x\in K$

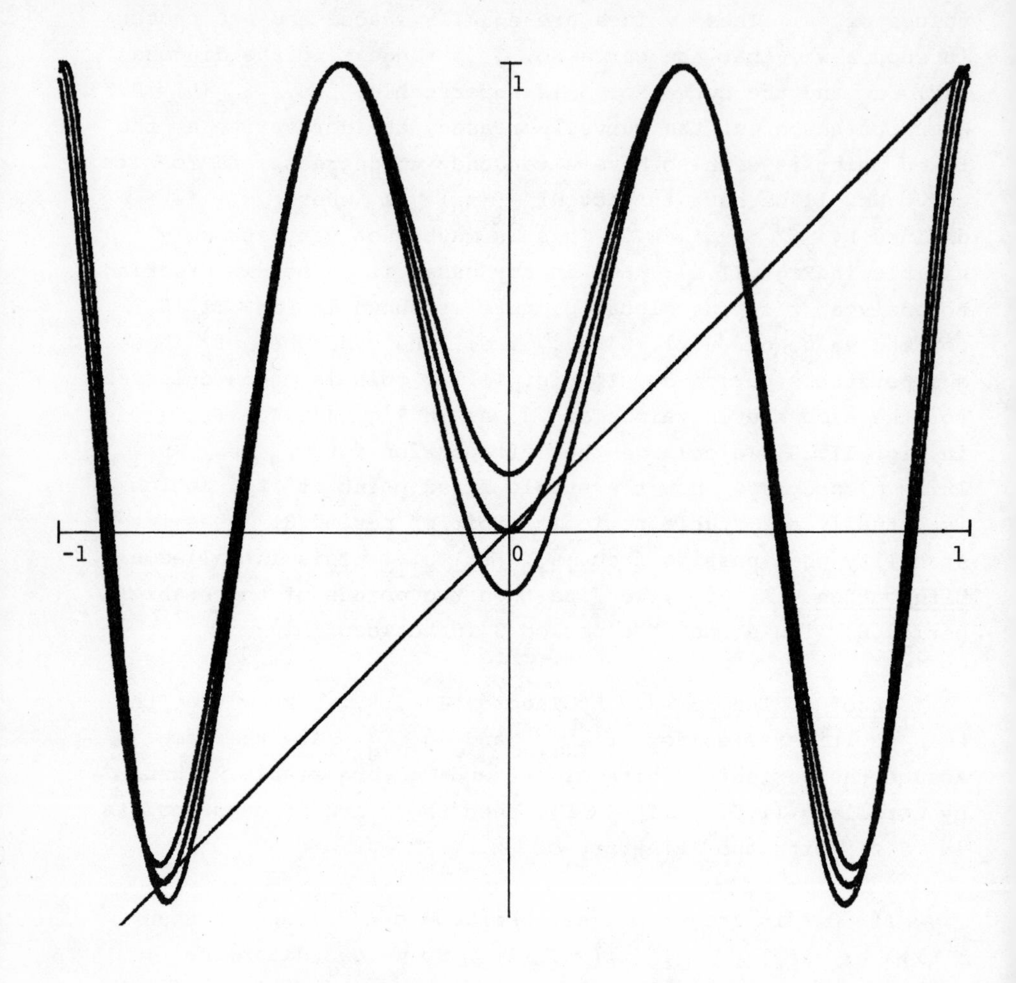

Fig. II.16. Graph of $f_\mu^3(x)$ for $f_\mu(x) = 1 - \mu x^2$ with three values of μ near 1.75487

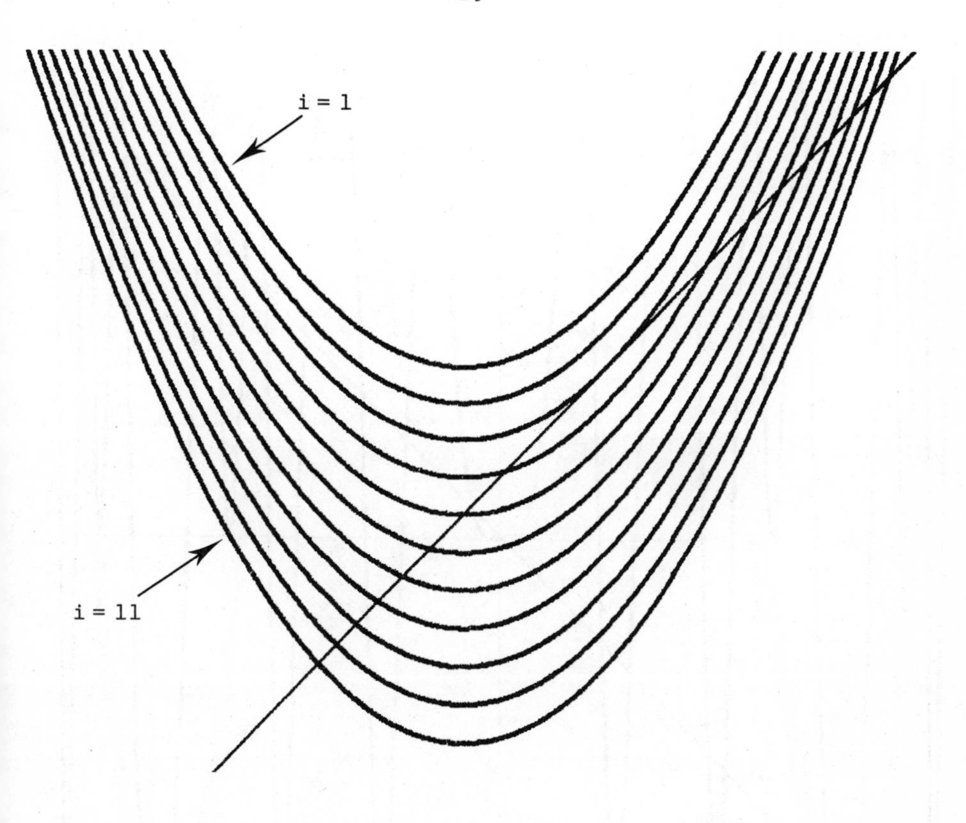

Fig. II.17. Graph of $f_{\mu}^{3}(x)$ for $f_{\mu}(x) = 1 - \mu x^2$ with $\mu_i = 1.7548 + (i-5)\cdot 0.0025$. Ten-fold enlargement w.r. to Fig. II.16 of region near $x = 0$.

i = 1,2 : no period 3

i = 3 : period 3, case of one-sided stability
$\underline{I}(1) = (RLR)^{\infty}$

i = 4 : stable period 3, case of two-sided stability
$\underline{I}(1) = (RLR)^{\infty}$

i = 5 : superstable period 3. $\underline{I}(1) = RLC$

i = 6 - 10: stable period 3, case of two-sided stability
$\underline{I}(1) = (RLL)^{\infty}$

i = 11 : stable period 6, period 3 is unstable. $\underline{I}(1) = (RLL)^{\infty}$

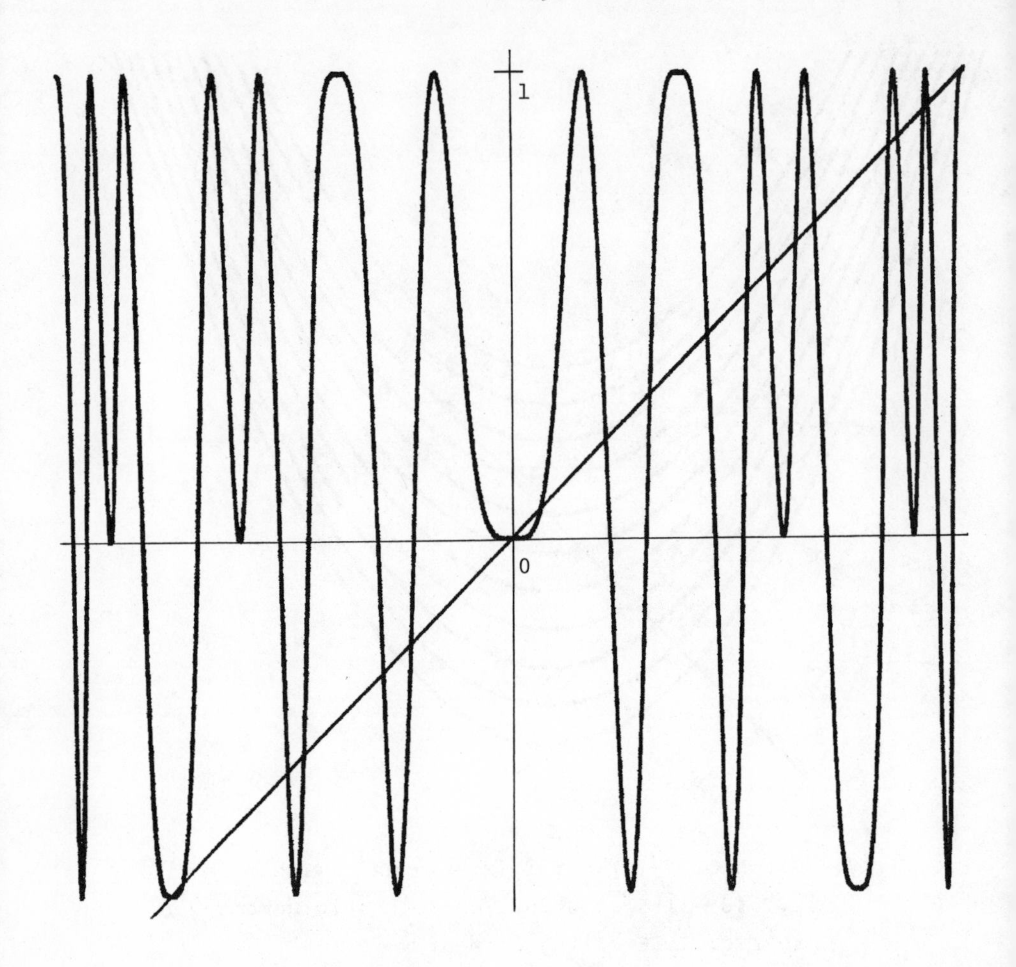

Fig. II.18. Graph of $f_\mu^6(x)$ where $f_\mu(x) = 1 - \mu x^2$ and $\mu = 1.7548$.

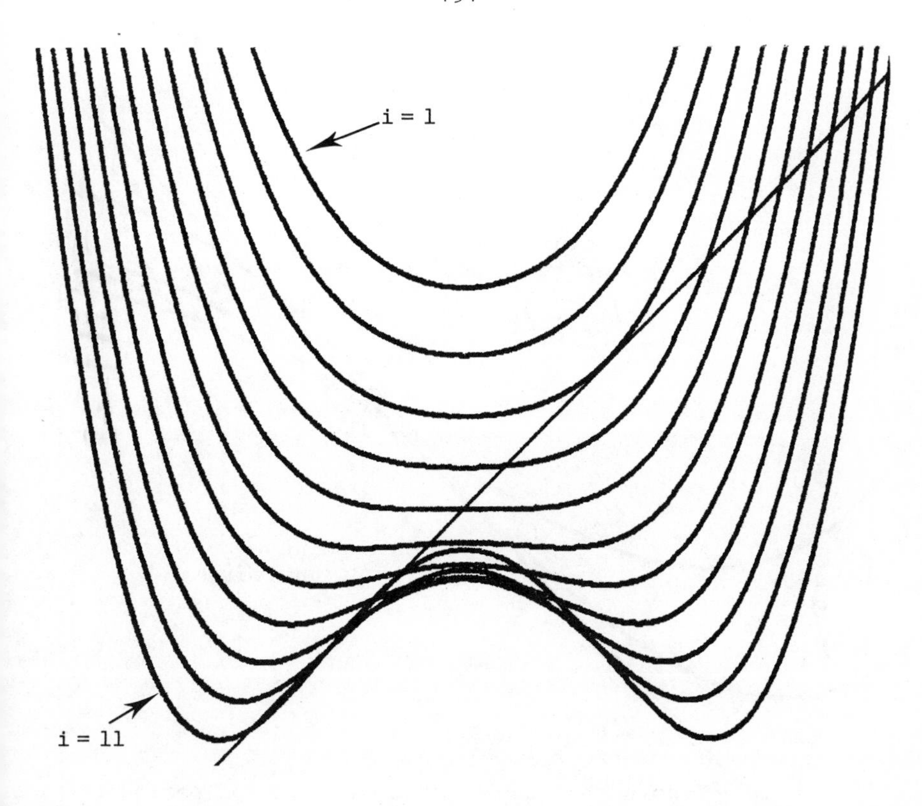

Fig. II.19. Graph of $f_{\mu}^{6}(x)$ for $f_{\mu}(x) = 1 - \mu x^2$ with $\mu_i = 1.7548 + (i-5) \cdot 0.0025$. Ten-fold enlargement w.r. to Fig. II.18 of region near $x = 0$.

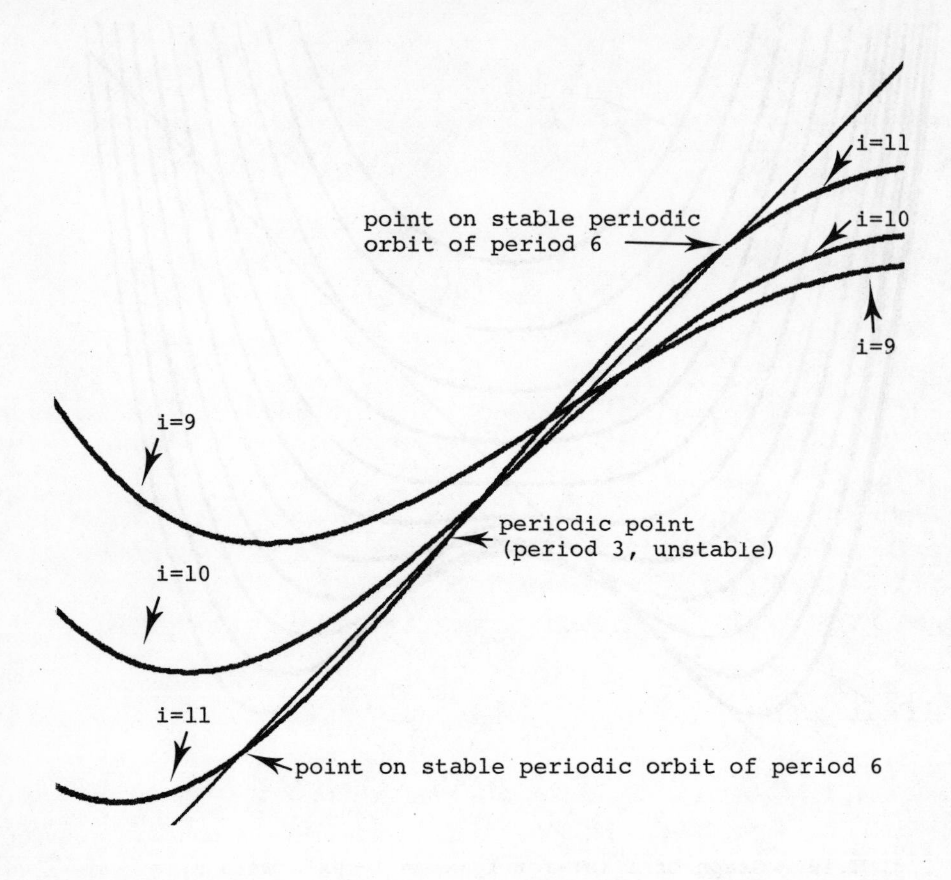

Fig. II.20. Detail of Fig. II.19 enlarged 100/3 w.r. to Fig. II.18. Center of figure is at $x = -100/3$.

by $h(f^j(x)) = g^j(h(x))$ and by the requirement that the itineraries $\underline{I}_f(x)$ and $\underline{I}_g(h(x))$ are the same (this identifies the component). It is easy to see that h, defined in this way, is a topological equivalence.

It remains to define h for $f|_{U_f}$. As we have seen in the proof of Proposition II.5.6, if U_f is as before, then if the periodic orbit is stable from both sides, U_f is of the form $(a;a')$, and one of the endpoints of U_f is periodic. Let $n>0$ be the smallest integer for which $f^n(U_f) \cap U_f \neq \emptyset$. By the definition of U_f, we must have $f^n(U_f) \subset U_f$, hence $f^n(a) = a$ or $f^n(a) = a'$. $f^n|_{U_f}$ is a map with one critical point since $x \in U_f$ implies $f^i(x) \neq 0$ for $i < n$. Since f^n has a fixed point p in U_f, it must be on the stable periodic orbit. There are different kneading sequences in the case when $Df^n(p)$ is positive, zero, and negative. We formulate now the topological equivalence for $F = f^n|_{U_f}$ and $G = g^n|_{U_g}$.

LEMMA II.6.4. <u>Assume F maps U_F into itself and has a maximum at c_F in U_F and a stable fixed point p_F in U_F, and suppose F is monotone on both sides of c_F. Similarly for G, U_G, c_G, p_G. Assume that if x is in the interior of U_F (resp. U_G), then $F^m(x) \to p_F$ (resp. $G^m(x) \to p_G$) as $m \to \infty$. Assume furthermore that $F'(p_F)$ and $G'(p_G)$ have the same sign or are both zero. Then F and G are topologically conjugate on U_F and U_G.</u>

REMARK. This lemma will be applied with $U_F = U_f$, $U_G = U_g$, $F = f^n|_{U_f}$, $G = g^n|_{U_g}$, or with $-f^n|_{U_f}(-.)$ if $f^n|_{U_f}$ has a minimum, and similarly for g.

<u>Proof</u>. The difficulty of the lemma is more in its formulation than in its proof, which is just a painstaking enumeration of the various cases. We present only the proof in the case of $F'(p_F) > 0$, this implies $p_F < c_F$. The interval $[p_F, c_F]$ has the property that $F[p_F, c_F] \subset p_F, c_F]$. Define h to be any strictly increasing function of $[F(c_F), c_F]$ to $[G(c_G), c_G]$. Then h extends to a homeomorphism of $[p_F, c_F]$ onto $[p_G, c_G]$ by the formula $h(F^m(x)) = G^m(h(x))$ and

$h(p_F) = p_G$. We extend now h to (p_F, p_F') by the formula $h(x') = (h(x))'$, where x' is defined by $F(x) = F(x')$, $x \neq x'$, and $h(x)'$ is defined by $G(h(x)') = G(h(x))$ and $h(x)' \neq h(x)$. In a similar way one treats the other situations of Lemma 5: If $p_F = c_F$ then one starts defining h from an interval $[x, F(x)]$ onto an interval $[y, G(y)]$ for arbitrary choices of $x \in U_F \setminus c_F$ and $y \in U_G \setminus c_G$ (on the same side of c_G as x is of c_F). If $F'(p_F) < 0$, one begins by defining h from the interval $[c_F, F^2(c_F)]$ onto $[c_G, G^2(c_G)]$. The extensions of h to topological equivalences are then uniquely determined. We leave the details to the reader.

Remarks and Bibliography. The question of topological classification of maps was of general concern for Milnor-Thurston [1977]. They were able to show that maps with identical itineraries are semiconjugate, i.e., the intertwining map is monotone, but not strictly monotone. The situation was then clarified through Guckenheimer's [1979] work. A very complete account for the case of diffeomorphisms of the circle can be found in Herman [1979]. Cf. also Parry [1964].

II.7 SENSITIVE DEPENDENCE ON INITIAL CONDITIONS

In this section, and the next, we analyze in more detail the S-unimodal maps without stable periodic orbit. We shall see later that such maps are quite frequent, and thus their study is not only justified from a mathematical point of view but is relevant as a somehow "physical" problem of one-dimensional dynamical systems.

Among the maps without stable periodic orbits different degrees of aperiodicity can be observed. The theory is very far from giving either a complete enumeration of these degrees, or establishing the exact relation between different aperiodic behaviors. Thus this is an interesting domain, in which a lot of open questions remain.

In this section, we discuss a notion of sensitive dependence on initial conditions, called _sensitivity_ for short. In some sense it is the weakest interesting form of aperiodic behavior. We say f has _sensitivity_ (_sensitive dependence on initial conditions_) if there is a set $Y \subset [-1,1]$ of positive Lebesgue measure and an $\varepsilon > 0$ such that for every $x \in Y$ and every neighborhood U of x, there is a $y \in U$ and an $n \geq 0$ such that $|f^n(x) - f^n(y)| > \varepsilon$.

Let us informally discuss this definition, which is due to Guckenheimer [1979]. The point is, of course, that U should be thought of as having diameter much less than ε so that the definition implies that x, y are guaranteed to be separated a lot (relatively speaking). In other words, no matter how small we choose U, at least two of its points are noticeably separated under the repeated action of f. This means that the orbit of x (under the repeated action of f) will depend in a sensitive way on the choice of the initial point x. Other definitions are of course possible. In particular, Ruelle [1979] has proposed to look for maps with an absolutely continuous invariant measure, for which $1/n \log|Df^n|$ tends to a strictly positive constant as $n \to \infty$, almost surely with respect to the invariant measure. This definition has the advantage of requiring more uniformity in the stretching than Guckenheimer's. Unfortunately, we have no results to offer in the direction of Ruelle's definition. The non-uniformity of Guckenheimer's definition comes from the fact that while $|f^n(x) - f^n(y)| > \varepsilon$ for some n, there may be a neighborhood W of x (with $y \notin W$) such that $f^n(x) - f^n(x') \sim 0$ for $x' \in W$.

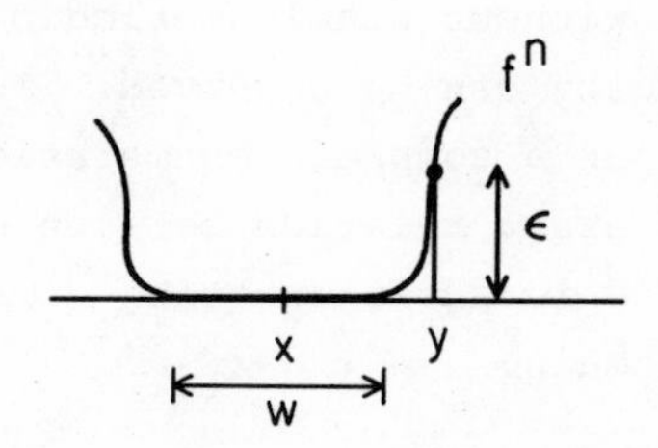

Figure II.21

Of course this seems highly implausible to be the case for all n, but we have no proof that it cannot occur.

After the discussion of the limitations of Guckenheimer's definition, we are now going to see its positive side. Namely it can be decided relatively easily, whether or not f has sensitivity. We first show that a map with stable periodic orbit does not have sensitivity.

THEOREM II.7.1. *If f is S-unimodal and has a stable periodic orbit of period k then f has no sensitivity.*

Proof. Let U be the stable neighborhood of 0. Every point x for which $f^n(x)$ is in the interior of U for some finite $n \geq 0$ has the property that for a sufficiently small neighborhood W of x all points $y \in W$ satisfy

$|f^m(y) - f^m(x)| < \varepsilon$ for all $m \geq 0$, since for $m \leq n$, $|f^m(y) - f^m(x)| < \sup_{z \in [-1,1]} |Df^m(z)| \cdot |x-y|$ while we can achieve $f^n(y) \in U$ for W sufficiently small, and $f^k|_U$ is a contraction. Hence the only points with possible sensitivity are those in $E_f = \{x : f^n(x) \notin U, \text{ for all } n \geq 0\}$. But $\lambda(E_f) = 0$ by Proposition II.5.7. Thus there is no sensitivity.

We shall now develop a criterion which will be useful to decide on sensitivity, and which in turn can be expressed through $\underline{K}(f) = \underline{I}_f(1)$ alone, as we shall see below.

We define x to be a <u>central fixed point</u> of f^n, $n > 1$ if (i) f^n is a homeomorphism on $(x;0)$ and (ii) $Df^n(x) > 0$. We say the central fixed point x of f^n is <u>restrictive</u>, if f^n maps $(x;x')$ into itself. Note that (i) implies that f^n is also a homeomorphism on $(0;x')$.

In particular we do not consider f^1 in the above definitions. The following figure illustrates f^n, $n > 1$ with a fixed point which is not central, central, and restrictive central, respectively.

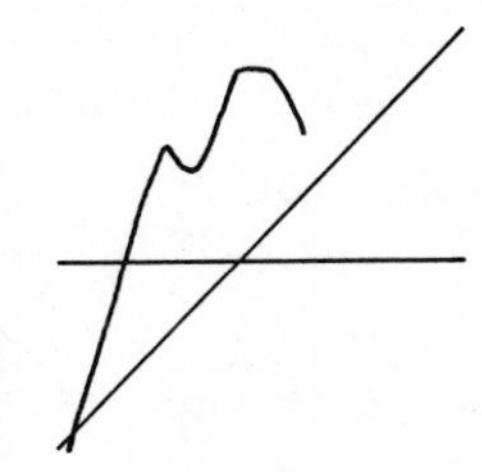

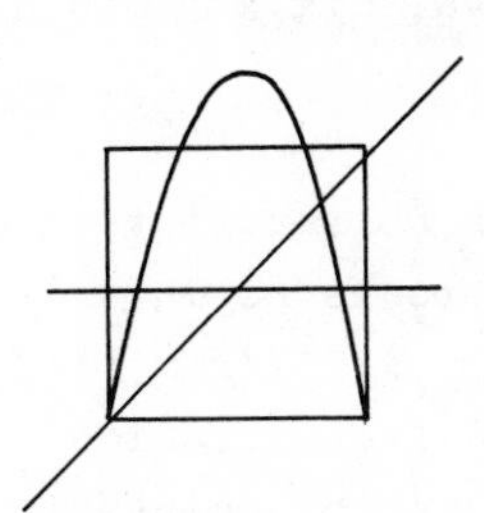

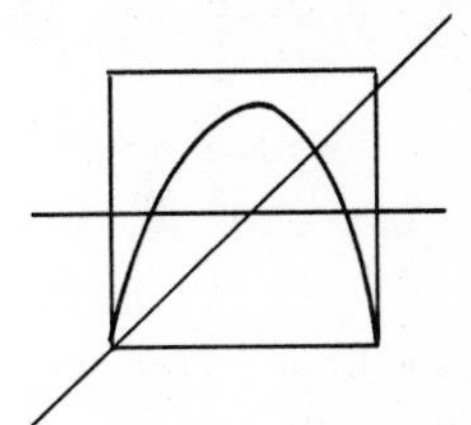

Figure II.22.

If x is a restrictive central point of period n and $U = (x;x')$, then $f^n(U) \subset U$. This implies that the set $\cup_{i=0}^{n-1} f^i(U)$ is invariant for f and that a point y which is mapped into this set cannot escape. Thus restrictive central points are barriers to the separation of points in orbits with nearby initial conditions.

We shall now mostly concentrate on functions f without stable periods. Then the definition above, which will be very useful for proofs, can be replaced by a simpler definition. We shall show below in which sense the two definitions are equivalent.

We shall say f has an <u>invariant segment</u> if for some $n \geq 1$, there is an interval (x,x') such that

(i) $f^n((x,x')) \subset (x,x')$,

(ii) f^n is monotone on $(x,0)$ and on $(0,x')$.

The relations between these definitions are described in the following

LEMMA II.7.2. *Let f be $\mathscr{C}^1$-unimodal without stable periodic orbit. f^n has no restrictive central point for all n if and only if f has no invariant segment.*

Proof. If f^n has a restrictive central point x for some n, then the interval $(x;x')$ is an invariant segment. This proves the if part.

Assume now f^n has no restrictive central point for all $n > 1$. Then we claim there is no invariant segment. It is useful to introduce $L_n = \{x : f^i((x;x')) \not\ni 0 \text{ for } 1 \leq i < n\}$. From the chain rule of differentiation, it is clear that the only zeros of $Df^n|_{\bar{L}_n}$ are the points in the boundary of L_n (L_n is an interval (y,y')) and 0.

We proceed now by induction on n. Note first that for any $\mathscr{C}^1$-unimodal map, $f(x,x') \not\subset (x,x')$ when $x \neq 0$, because $f(0) = 1$ and (x,x') is open in $[-1,1]$. Thus the case $n = 1$ is handled. Assume now $n > 1$ and that we have shown for all i, $1 \leq i < n$ that $f^i((y;y')) \not\subset (y;y')$ for all $y \in L_i$. We wish to show that the same is true for $i = n$. There are two cases. In the first case, f^n has a central fixed point, x. By assumption, this fixed point is not restrictive. Throughout the remainder of the argument, we shall assume for definiteness $x < 0$. Then $f^n(0) > x'$ and $Df^n(x) > 1$ since f^n has no stable periodic orbit. We want to show that for $y \in L_n$, $y < 0$ one has $f^n((y,y')) \not\subset (y,y')$. If $0 > y > x$, then $(y,y') \subset (x,x')$ and $f^n((y,y')) \ni f^n(0) > x' > y'$ and thus the assertion follows in this case. If $y < x$ and $y \in L_n$, then we must have $f^n(y) < y$, since $Df^n(z) > 0$ for all $z \in L_n$ with $z \leq x$ and since f^n has no stable fixed point. Hence $f^n((y,y')) \not\subset (y,y')$ in this case too. This completes the argument in the first case.

In the second case, f^n has no central fixed point. We still assume f^n is increasing on $\{x: x \in L_n$ and $x < 0\}$. Let $z < 0$ be the left endpoint of L_n. (Assume for simplicity that $f(-1) < 0$, otherwise extend f and the interval on the left in order to achieve this condition, and then rescale coordinates. This does not change the absence of stable periodic orbits). Then $Df^n(z) = 0$ when $n > 1$ and hence we must have $f^i(z) = 0$ for some i, $1 \leq i < n$. It follows that $f^n(z) = f^{n-i}(0)$. If $f^{n-i}(0) \notin L_n$, we find $f^n(z) \notin L_n$, and the only way this is possible from the monotonicity of f^n is $f^n(z) \leq z$. If $f^n((y;y')) \subset (y;y')$ for some $y \in (z;z')$ we arrive at a contradiction with the assumption that f^n has no restrictive central fixed point. If $f^{n-i}(0) \in L_n$ and $f^{n-i}(0) > 0$ we must have from the monotonicity of $f^n|_{(z,0)}$ and from the recursive assumption that $f^{n-i}|_{(z,z')} \geq f^{n-i}(0)$. We must then have $f^{n-i}(z) > z'$ because otherwise $f^{n-i}((z,z')) \subset (z,z')$ which would contradict the induction hypothesis. By the same argument using $f^i(z) = 0$ we find $f^i(0) > z'$ (or $f^i(0) < z$) so that, by continuity, $f^n(w) = f^{n-i}(f^i(w)) > z'$ for some w in L_n. Hence $f^n(L_n) \not\subset L_n$, and thus $f^n((y,y')) \not\subset (y,y')$ for every $y < 0$, $y \in L_n$. If $f^{n-i}(0) < 0$, we must still have $f^{n-i}|_{(z,0)} \leq f^{n-i}(0)$. Now we can argue as before. In fact, an extension of the argument implies $f^{n-i}(L_n) \ni 0$, a contradiction. Thus the assertion follows in all cases.

In the same spirit, we can prove the following two lemmas, which will be used below.

LEMMA II.7.3. *If* f *is* S-*unimodal and has no stable periodic orbit, then for every neighborhood* U *of* 0 *and every* N *there is an* $n \geq N$ *such that* f^n *has a central fixed point in* U.

Proof of Lemma II.7.3. By Proposition II.5.5, the set of pre-images of 0 is dense in $[-1,1]$. Therefore, given $N > 0$ and a neighborhood U of 0 we can find $n > N$ such that the critical point x of f^{n+1} closest to 0 (but different from it) satisfies $f^n(x) = 0$ and $x \in U$. Then f^n has a fixed point q in one of the intervals $(x;0)$ or $(x';0)$, because $f^n(x) - x$ and $f^n(x') - x'$ have opposite signs, and $f^n(0) \neq 0$. We do not expect $Df^n(q) > 0$, in general. Let $K_n = \{y: y \in S, f^i(y) \notin (y;y'), i < n, f^n(y) \in (y;y')\}$, where S is the interval of monotonicity of f^n containing x. The ends of S are not in K_n, but the points in the boundary of K_n satisfy one of the equations $f^n(y) = y$ or $f^n(y) = y'$ by Lemma II.5.6. The points at opposite ends of an interval in K_n do not both satisfy the same equation because f has no stable periodic orbits. This implies that in K_n and in K_n' (defined through the usual relation) one of the ends z satisfies $f^n(z) = z$. At one of them $Df^n(z) > 0$ and this point is a central fixed point for f^n. This proves Lemma 3.

LEMMA II.7.4. *Let* p *be a restrictive central point for* f^n. *Assume* f *has no stable periodic orbit*. *Then* $0 \in f^n((p;p'))$.

Proof. Assume for example $p < 0$, hence $Df^n(p) > 0$. From $Df^n(p) > 1$, there is a unique point q in $(p,p']$ such that $f^n(q) = q$. If $f^n(y) < 0$ for all y in $[p,p']$, then $f^n(q) = q < 0$. Since $Df^n(q) > 1$, $Df^n(p) > 1$, we have $Df^n(y) > 1$ for all $y \in [p,q]$, since $|Df^n|$ has no non-zero minimum on $[p,q]$. Therefore $f^n(q) = p + \int_p^q Df^n(x)\,dx > q$, a contradiction.

LEMMA II.7.5. $f^u(p) \notin (p,p']$ <u>for any</u> $u \le n$.

<u>Proof</u>. The case $u = n$ is obvious. To prove the other cases, assume the contrary, i.e., $f^u(p) \in (p,p']$ for some $u < n$. Let q be the unique point in $(p,p']$ such that $f^n(q) = q$. Since $f^u(p)$ is a point of the orbit of p, it is a fixed point of f^n. Therefore $f^u(p) = q$, i.e., q is on the orbit of p. But this is impossible since $Df^n(q) \ne Df^n(p)$ (they have opposite sign and are not equal to zero). This proves the lemma.

We state and prove three auxiliary lemmas.

LEMMA II.7.6. <u>Suppose</u> f <u>has no stable periodic orbit and suppose</u> f^k <u>has no restrictive central point for</u> $k > n$ <u>and that</u> p <u>is central for</u> f^n. <u>Then there is a central point</u> q <u>and a</u> k <u>such that</u> $q \in (p;p')$ <u>but</u> $f^k((q;q')) \supset (p;p')$.

<u>Proof of Lemma II.7.6</u>. For simplicity, we shall restrict ourselves to the case $p < 0$, the case $p > 0$ is similar. Let $\bar{p}$ be the fixed point of f^n contained in (p,p'), and denote by I the interval $[\bar{p}',\bar{p}]$. Let k_1 be the smallest integer such that $0 \in f^{k_1}([\bar{p}',\bar{p}])$ (such a k_1 exists by Corollary II.5.5 since f has no stable periodic orbit). Let y be the leftmost point in $(0,\bar{p}]$ such that $f^{k_1}(y) = 0$. We now look at the map $f^{k_1}|_{(0,\bar{p}]}$. Since there is no pre-image of zero of order less than k_1 on $(0,\bar{p}]$, $f^{k_1}|_{(0,\bar{p}]}$ is monotone. We now claim that $f^{k_1}(\bar{p}) \notin (\bar{p}',\bar{p})$. First of all, we have $k_1 > n$ ($k_1 \ge n$ since f^n is monotone on $(0,\bar{p}]$, and $k_1 \ne n$ since $0 \notin f^n([\bar{p}',\bar{p}])$). We can write $k_1 = rn + s$ with $0 \le s < n$. Assume first $f^{k_1}(\bar{p}) \in (\bar{p}',\bar{p})$ with $s \ne 0$. We have

$f^{k_1}(\bar{p}) = f^s(f^{rn}(\bar{p})) = f^s(\bar{p}) \in (\bar{p}',\bar{p})$. However $f^n(f^s(\bar{p})) = f^s(f^n(\bar{p})) = f^s(\bar{p})$, and $f^s(\bar{p})$ is a fixed point of f^n contained in $(\bar{p}',\bar{p})$, a contradiction with the construction of $\bar{p}$. The case $s = 0$ gives

$$f^{k_1}(\bar{p}) = f^{rn}(\bar{p}) = \bar{p} \in (\bar{p}',\bar{p})$$

which is obviously impossible.

From the monotonicity of f^{k_1} on $[\bar{p}',0]$, and on $[0,\bar{p}]$ and the facts that $f^{k_1}(y) = 0$ and $f^{k_1}(\bar{p}) \notin (\bar{p}',\bar{p})$, we derive the existence of a central point q for f^{k_1}, $q \in (\bar{p}',\bar{p})$. Since $k_1 > n$, q is not restrictive.
Therefore $[q';q) \subset f^{k_1}((q;q'))$. Some iterate then satisfies

$$f^{mk_1}((q;q')) \supset [\bar{p}';q) \quad (\text{or } \supset [\bar{p};q)) \quad .$$

Pursuing only the first alternative, the monotonicity of f^n implies $f^{mk_1+sn}((q;q')) \supset (p,p')$ for some s, since p is an unstable fixed point.

LEMMA II.7.7. _Assume_ f _has no stable periodic orbit, let_ p_1 _be a restrictive central point for_ f^n _and_ p_2 _a restrictive central point for_ f^m, $m > n$. _Then_ $[p_2;p_2'] \subset (p_1;p_1')$.

Proof. Assume $(p_1;p_1') \subset (p_2;p_2')$. From Lemma 4 we have $0 \in f^n((p_1;p_1'))$, therefore $Df^m(y) = 0$ for some $y \neq 0$ in $(p_2;p_2')$, a contradiction.

LEMMA II.7.8. _Assume_ f _has no stable periodic orbit. Let_ p _be restrictive and central for_ f^n. _Then_

$$f^i(\overline{(p;p')}) \cap \overline{(p;p')} = \emptyset \quad \text{\underline{for}} \quad i = 1,2,\ldots,n-1.$$

Proof. Assume $f^i(\overline{(p;p')}) \cap \overline{(p;p')} \neq \emptyset$ for some $i < n$. There is then a y in $(p;p')$ such that $f^i(y) = p^\#$. Therefore $f^n(y) = f^{n-i}(p^\#) = f^{n-i}(p)$. By Lemma 5 $f^{n-i}(p) \notin [p,p']$. Thus $f^n(y) \notin [p,p']$, which contradicts the fact that p is restrictive, central.

THEOREM II.7.9. (Guckenheimer [1979].) *Let f be S-unimodal and suppose f has no stable periodic orbit. Then f has sensitivity to initial conditions if and only if there is an N such that for all $n \geq N$, f^n does not have a restrictive central point.*

Proof. "Only if" part. Suppose f is S-unimodal and has no stable periodic orbit. Suppose furthermore that f has infinitely many restrictive central points $p_1, p_2, \ldots, p_k, \ldots$, with periods $n_1 < n_2 < n_3 < \ldots < n_k < \ldots$, respectively. Set $U_k = [p_k; p_k']$, and notice that $U_{k+1} \subset U_k$ by Lemma 7. For almost all $x \in [-1,1]$, there is an N such that

$$f^n(x) \in \bigcup_{i=0}^{n_k - 1} f^i(U_k) \equiv S_k$$

for $n \geq N$ by Theorem II.5.2.3. It follows that there is a set $K \subset [-1,1]$ of full Lebesgue measure such that for $x \in K$, $\omega(x) = \cap_{n\geq 0}(\{f^i(x);\ i \geq n\}^-)$ is contained in $\Lambda = \cap_k S_k$. $\omega(x)$ is called the ω-*limit set* of x.

We first show that Λ contains no non-trivial interval. We claim that $p_k \to 0$ if $k \to \infty$. Assume the opposite, i.e., for some $\eta > 0$, $\eta \in (p_k; p_k')$ for all k. By Lemma 3, there is an n such that f^n has a central fixed point p in $(\eta;\eta')$. Since p is not restrictive, we have $0 \in f^n((\eta;\eta'))$, which contradicts the existence of a restrictive central point with period greater than n. Now assume Λ contains a non-trivial interval I. By Cor. II.5.5, I contains a preimage x of zero, i.e., $f^s(x) = 0$ for some s and some x in I. Since Λ is f-invariant, we have $0 \in f^s(I) \subset \Lambda$. Moreover, $J = f^s(I)$ is also a non-trivial interval. From

Lemma 8 and $f^s(I) \subset S_k$ we deduce $f^s(I) \subset U_k$ for all k, which contradicts $p_k \to 0$ if $k \to \infty$. We next show that $\mu_k = \max\{\lambda(f^i(U_k)): i = 0,1,\ldots,n_k-1\} \to 0$ as $k \to \infty$. Assume this is not true, i.e., there is an $\eta > 0$, such that $\mu_k > \eta$ for all k, which implies that for all k, there is an i, $0 \le i < n_k - 1$ with $\lambda(f^i(U_k)) \ge \eta$. We claim there is then a sequence of indices i_k, $0 \le i_k < n_k$ such that

(i) $f^{i_k}(U_k) \supset f^{i_{k+1}}(U_{k+1})$ for every k,

(ii) $\lambda(f^{i_k}(U_k)) \ge \eta$ for every k.

From this it is easy to derive the result. Let $I = \cap_k f^{i_k}(U_k)$, we have $\lambda(I) \ge \eta$, $I \subset \Lambda$, and I is a non-trivial segment, a contradiction. We now construct a sequence i_k. Let $\Omega_k = \{i: 0 \le i < n_k, \lambda(f^i(U_k)) \ge \eta\}$.

From our assumption, we have $\Omega_k \ne \emptyset$. We claim that in Ω_0 there is an i_0 such that for every k there is an i in Ω_k with $f^i(U_k) \subset f^{i_0}(U_0)$ (if this is not true, there is a k such that for all i,

$$f^i(U_k) \not\subset \bigcup_{m=0}^{n_0-1} f^m(U_0)$$

or $\lambda(f^i(U_k)) < \eta$ which is impossible). If there are different possible i_0, we choose the smallest one. Assume now $i_0,\ldots,i_s$ have been chosen, and that $f^{i_s}(U_s)$ is such that for any $m \ge s$, there is an i in Ω_m with $f^i(U_m) \subset f^{i_s}(U_s)$. Let $\Omega'_{s+1} = \{i \in \Omega_{s+1}: f^i(U_{s+1}) \subset f^{i_s}(U_s)\}$. Then, there is a j in Ω'_{s+1} such that for all $m > s+1$, there is an i in Ω_m with $f^i(U_m) \subset f^j(U_{s+1})$ (if not, we have a contradiction with a property of i_s). We now choose for i_{s+1} the smallest j in Ω'_{s+1} with the above property. This defines recursively the sequence $\{i_k\}_{k=0,1,\ldots}$ and (i) and (ii) are easy to verify.

We now notice that $x \in K$ implies that $f^N(x) \in U_k$ for some N. We can find a neighborhood V of x such that $f^N(V) \subset U_k$ and $\max_{i \leq N} \lambda(f^i(V)) < \varepsilon$. Then for all n we have $\lambda(f^n(V)) < \varepsilon$ since $\lambda(f^i(U_k)) \leq \mu_k < \varepsilon$ for k sufficiently large. This shows f does not have sensitivity.

<u>"If"</u> part. Consider a S-unimodal f without stable periodic orbit and f^n without restrictive central point when $n > N$. We shall show that there is an $\varepsilon > 0$ such that for every non-trivial interval J, there is an n with $\lambda(f^n(J)) > \varepsilon$. This obviously implies sensitivity. Since the set of pre-images of 0 is dense by Corollary II.5.5, we may assume J contains 0. Given $q_0 = p$ a central point for f^N, we can, using Lemma 6, find a sequence $\{q_k\}$ of central points and integers n_k, such that $f^{n_k}((q_k;q_k')) \supset (q_{k-1};q_{k-1}')$ and q_k is closer to 0 than q_{k-1}. We now show that $q_k \to 0$. If this is not true, we have $\eta > 0, \eta \in (q_k;q_k')$ for all k and some η. By Lemma 3, there is a central point in $(\eta;\eta')$ of period n greater than N. Therefore this point is not restrictive by hypothesis, and we obtain $0 \in f^n((\eta;\eta'))$, a contradiction. By construction,

$$f^{n_k+n_{k-1}+\ldots+n_1}((q_k;q_k')) \supset (p;p') \quad .$$

Hence $f^n(J) \supset (p;p')$ for some n by the preceding construction. Setting $\varepsilon = \lambda((r;r'))$, where r is a central point for f^{n_1}, we see that for every interval J there is a k with $\lambda(f^k(J)) > \varepsilon$. This proves the theorem.

We would like to compare the maps with sensitivity to some simple maps, namely to the piecewise linear maps $g_\mu(x) = 1 - \mu|x|$. This desire, to bring maps to an easy form has a long history for the diffeomorphisms of the circle. Not every diffeomorphism of the circle is topologically conjugate to a rotation, but among the $\mathscr{C}^3$ diffeomorphisms, those with sufficiently irrational rotation numbers are (see Hermann [1979]). An analogous fact is true for S-unimodal maps, namely those with "sufficiently nonstable" periodic orbits will be seen to be topologically conjugate to piecewise linear maps. Let us start with an analysis of piecewise linear maps.

LEMMA II.7.10. If $\sqrt{2} < \mu \leq 2$, then g_μ has no restrictive central points.

Proof. Let p be a central fixed point of g^k, $k > 1$. Then $p' = 1 - p$ and g^k is monotone on $(p;0)$. By assumption, $|Dg^k(x)| > 2$, when $x \neq 0$. Therefore $|g^k(0) - g^k(p)| > 2p$. But $|p-p'| = 2|p|$, this implies $p' \in (p;g^k(0))$, and p is not restrictive.

This has an obvious consequence:

COROLLARY II.7.11. If f is S-unimodal and has a restrictive central point, then it cannot be topologically conjugate to a g_μ with $\sqrt{2} < \mu \leq 2$.

Using the notion of growth number, defined as $\limsup_{k\to\infty} (N_k)^{1/k}$, where N_k is the number of fixed points of f^k, one can be more specific about the relation between f and g_μ. (The growth number of g_μ is μ.) We shall not expand on this subject, and its connection with topological entropy. See end of Section 8.

A positive criterion for deciding on conjugacy is given by the following

THEOREM II.7.12. Let f be S-unimodal and have no stable periodic orbit and no restrictive central point. Then there is a $\mu \in (\sqrt{2},2]$ such that $f|_{J(f)}$ is topologically equivalent to $g_\mu|_{[1-\mu,1]}$.

Proof. Since f has no stable periodic orbit, the set $\{x\colon f^n(x) = 0 \text{ for some } n \geq 0\}$ is dense, by Corollary II.5.5.

If we can show that there is a g_μ such that its kneading sequence equals $\underline{K}(f)$ then by the proof of Theorem II.6.1, $f|_{J(f)}$ and $g_\mu|_{[1-\mu,1]}$ are topologically conjugate. This will be done in Lemmas III.1,4, 6, by a method similar to the proof of Lemma II.3.9. We show there that <u>every maximal</u> $\underline{A} > RLR^\infty$, not of the form of $\underline{A} = \underline{B}*\underline{Q}$, $\underline{Q} \neq C$, occurs as a $\underline{K}(g_\mu)$. To complete the proof of the theorem, we will show in Corollary 14, that unimodal maps, having no restrictive central point and no stable periodic orbit, are characterized by itineraries which are never of the form $\underline{B}*\underline{Q}$. This obviously proves the theorem.

LEMMA II.7.13. *Let* f *be S-unimodal. Then* f *has an invariant segment if and only if* $\underline{K}(f) = \underline{B}*\underline{Q}$, *for some* $\underline{B} \neq \emptyset$, $\underline{Q} \neq C$.

Proof. If f has an invariant segment, then f^n maps (x,x') into itself for some $n > 1$ and some $x < 0$. Define $J_i = f^i((x,x'))$ for $i = 1,2,\ldots,n-1$. By the definition of invariant segment $0 \notin J_i$. Hence the itinerary $\underline{I}_f(1)$ must be of the form $\underline{I}_f(1) = \underline{B}Q_0\underline{B}Q_1\ldots$, where $B_i = R$ if J_i is to the right of 0, L otherwise. This proves the only if part. Assume now $\underline{I}_f(1) = \underline{B}*\underline{Q}$. Denote by J the interval $[f^n(1),1]$ where $n = |\underline{B}C|$. Note that $\underline{I}_f(f^n(1)) = \underline{B}*\mathscr{S}\underline{Q}$. Let $\hat{J}$ be the set of points with itineraries $\underline{A}$ satisfying $\underline{B}*\mathscr{S}\underline{Q} \le \underline{A} \le \underline{B}*\underline{Q}$. This is an interval containing J. By Lemma II.1.3 and Theorem II.2.7, we see from the maximality of $\underline{Q}$ that f^n maps $\hat{J}$ into itself. Furthermore, for every $x \in \hat{J}$ $\underline{I}_f(x) = \underline{B}X_0\underline{B}X_1\ldots$, and hence $f^n|_{\hat{J}}$ has no other critical points than those of the map $f|_{f^{-1}(\hat{J})}$. Defining $(x,x') =$ interior $f^{-1}(\hat{J})$, we complete the proof of the lemma.

Our preceding discussion leads by combination of Lemmas 2 and 13 to the following.

COROLLARY II.7.14. *Let* f *be S-unimodal without stable periodic orbit* (i.e., $\underline{K}(f)$ *is infinite and not periodic by*

Proposition II.6.2). Then f has a restrictive central point if and only if $\underline{K}(f) = \underline{B}*\underline{Q}$ for some $\underline{B} \neq \emptyset$ and $\underline{Q} \neq C$.

Remarks and Bibliography. The notion of sensitivity to initial conditions has been used in many instances, but the precise definitions vary from reference to reference. In ergodic theory, Walters [1975(1)] uses the word expansiveness but requires sensitivity for every pair of points. This is useful for maps on the circle, but too strong for maps on the interval, since there may be very long periodic orbits with even period $2n$ for which $f^i(x)$ and $f^{i+n}(x)$ stay close for all i, when x is on the periodic orbit.

The main merit of Guckenheimer's [1979] definition is its good applicability. In a way, it has a precursor in the work of Li-Yorke [1975], cf. also Remarks and Bibliography for Section 3.

The definition of Ruelle [1978(1)] is the most relevant one from a probabilistic point of view, but unfortunately, few good results are known so far. Most numerical experimentalists take the Liapunov characteristic exponent (Ruelle [1978(2)]) as a test for sensitivity. In practice, this means that one calculates $\log[Df^n(x)]/n$ for large n. Nice examples are given by Shaw [1978] (cf. Section I.5) for maps $x \to 1 - \mu x^2$ and by Feit [1978] for the Hénon maps in $\mathbb{R}^2$:

$$\begin{bmatrix} x \\ y \end{bmatrix} \to \begin{bmatrix} 1 - \mu x^2 + y \\ 0.3x \end{bmatrix}$$

The only a priori bound is the growth number (Milnor-Thurston [1977]), defined as $\lim_{n\to\infty} [\log N_n]/n$, where N_n is the lap number of f^n, i.e., the number of monotone pieces of f^n (= 1 + number of critical points of f^n). This is an upper bound for the Liapunov exponent, Ruelle [1978(2)]. See also end of Section 8.

II.8 ERGODIC PROPERTIES

In this section, we present a part of the beautiful theory of Misiurewicz [1980] on the existence of an invariant measure which is absolutely continuous with respect to Lebesgue measure and which is ergodic, for certain S-unimodal maps. We assume some familiarity with general ergodic theory.

After introducing some terminology, we state the results. We then give a partial interpretation of their meaning, which is complementary to the more intuitive description in Section II.2, which is sometimes based on non-proved extrapolations from established facts.

We consider only measures on the usual Borel-σ-algebra. A measure ν on $[-1,1]$ will be called an invariant measure for f if for every measurable subset E of $[-1,1]$ one has $\nu(E) = \nu(f^{-1}(E))$. We shall only consider probability measures on $[-1,1]$, i.e., $\nu([-1,1]) = 1$. Among the measures on $[-1,1]$, those which are absolutely continuous with respect to Lebesgue measure will play an important role. We shall for simplicity speak about absolutely continuous measures for short, and reserve the symbol λ for the Lebesgue measure itself. A measure ν is absolutely continuous if $d\nu/d\lambda$ is defined and is in $L_1(\lambda)$.

The easiest example of a unimodal map with absolutely continuous invariant measure is given by $f(x) = 1 - 2|x|$. It is easy to see from the definition that λ itself is an invariant measure. Every interval $I \subset [-1,1]$ has two pre-images $f_+^{-1}(I)$ and $f_-^{-1}(I)$ to the left and to the right of zero, and the length of $f_\pm^{-1}(I)$ is exactly 1/2 the length of I.

There is a quite general criterion for the existence of absolutely continuous measures for maps which are _everywhere expanding_. We define a map to be everywhere expanding if

(i) There is a finite set $x_0 = -1 < x_1 < x_2 < \ldots < x_n < 1 = x_{n+1}$.

(ii) For all $i = 0,\ldots,n$, f is $\mathscr{C}^2$ on (x_i, x_{i+1}) and can be extended as a $\mathscr{C}^2$ function to $[x_i, x_{i+1}]$.

(iii) $|f'|\big|_{(x_i,x_{i+1})} > \alpha > 2$.

Then one has the

THEOREM II.8.1. (Lasota-Yorke [1973]). _If_ f _is everywhere expanding, then_ f _has an absolutely continuous invariant measure_.

We shall take the proof of this theorem as an occasion to develop some intuitive ideas about existence of invariant measures which will be useful in understanding the results when f is not everywhere expanding. (The assumption $\alpha > 2$ is too strong and could be replaced by $\alpha > 1$. One considers then first f^N instead of f, with N such that $\alpha^N > 2$. Since N is finite, it is then relatively straightforward to transfer the invariant measure back to f.)

Assume for the moment that we have an absolutely continuous invariant measure ν for f whose density is the function h, i.e., $\nu(dx) = h(x)dx$. Then it is easy to see that h satisfies the equation

$$h(x) = \sum_{y \in f^{-1}(x)} \frac{h(y)}{|f'(y)|}$$

at those points where this expression makes sense. It is thus useful to consider the Perron-Frobenius operator f_* from the set of measurable functions h to itself, defined by

$$f_*(h)\cdot\lambda = f^*(h\cdot\lambda)$$

where $(f^*(\mu))(E) = \mu(f^{-1}(E))$. These two equations imply the more readable form

$$\int_{f^{-1}(E)} \phi \, d\lambda = \int_E f_*(\phi) \, d\lambda \quad .$$

Note that $(f^n)_* = (f_*)^n$, so that the parentheses are in fact superfluous. An absolutely continuous invariant measure for f has a density h which is in $L_1(\lambda)$ and which is a fixed point of f_*. So the problem of proving existence of absolutely continuous invariant measures can be brought to a fixed point problem for the operator f_*. It is clear that imposing $|f'| > \alpha > 2$ will help a lot in making f_* a "nice" operator, and the whole problem for $\mathscr{C}^1$-unimodal maps will come from the singularity of $|f'|^{-1}$ at $x = 0$. Let us proceed now to the proof of Theorem 1. We denote by v_a^b the variation of f over the closed interval $[a,b]$. We claim that

LEMMA II.8.2. (i) <u>For every</u> $h \in L_1$, <u>the sequence</u>

$$\frac{1}{n} \sum_{k=0}^{n-1} f_*^n h$$

<u>is convergent in</u> L_1 <u>and its limiting function</u> h^* <u>is in</u> L_1.

(ii) $h^*\cdot d\lambda$ <u>is an invariant measure for</u> f. h^* <u>is of bounded variation and there exists a constant</u> c <u>depending only on</u> f <u>such that</u>

(iii) $$\bigvee_{-1}^{1} h \le c \, \|h\|_{L_1}$$

(iv) <u>If</u> $h \geq 0$ <u>a.e. then</u> $h^* \geq 0$ <u>a.e.</u>

(v) $$\int_{-1}^{1} h \, d\lambda = \int_{-1}^{1} h^* \, d\lambda.$$

The proof of these statements is mostly computational. It is useful to introduce $f_i = f|_{(x_i, x_{i+1})}$, $i = 0, \ldots, n$ and $\phi_i = f_i^{-1}$. Then $|\phi_i'| \leq \alpha^{-1}$. We consider now f_* as a map from the set of measurable functions to itself.

Note that with $\sigma_i = |\phi_i'|$, we have

$$(f_* h)(x) = \sum_{i=0}^{n} h(\phi_i(x)) \sigma_i(x) \chi_i(x) ,$$

where χ_i is the characteristic function of the interval $J_i = \phi_i([x_i, x_{i+1}])$. Let now h be a function of bounded variation over $[-1,1]$. Then we have

$$\bigvee_{-1}^{1} f_* h \leq \sum_{i=0}^{n} \bigvee_{J_i} h \circ \phi_i \cdot \sigma_i + \alpha^{-1} \sum_{i=0}^{n} (|h(x_i)| + |h(x_{i+1})|) \qquad (*)$$

Let us bound the first sum, by the chain rule for variation by

$$\begin{aligned} \bigvee_{J_i} h \circ \phi_i \cdot \sigma_i &= \int_{J_i} |d(h \circ \phi_i \cdot \sigma_i)| \\ &\leq \int_{J_i} |h \circ \phi_i| |\sigma_i'| \, d\lambda + \int_{J_i} \sigma_i |d(h \circ \phi_i)| \\ &\leq K \int_{J_i} |h \circ \phi_i| \sigma_i \, d\lambda + \alpha^{-1} \int_{J_i} d(h \circ \phi_i) \quad , \end{aligned}$$

where

$$K = \sup_{i,x} |\sigma_i'| / \inf \sigma_i$$

and this quantity is bounded by our assumptions on f. Changing variables we obtain

$$\bigvee_{J_i} h\circ\phi_i|\phi_i'| \le K\int_{x_i}^{x_{i+1}} |h|\, d\lambda + \alpha^{-1}\int_{x_i}^{x_{i+1}} |dh| \quad .$$

Next we bound the second term in (*). Let $d_i = \inf\{|h(x)|: x \in [x_i, x_{i+1}]\}$. Then we have

$$|h(x_i)| + |h(x_{i+1})| \le \bigvee_{x_i}^{x_{i+1}} h + 2d_i \quad ,$$

but also

$$d_i \le \Delta^{-1}\int_{x_i}^{x_{i+1}} |h|\, d\lambda \quad ,$$

where $\Delta = \inf_i |x_{i+1} - x_i|$. Adding the various terms together we find

$$\bigvee_{-1}^{1} f_* h \le (K+2\Delta^{-1})\, ||h||_{L_1} + 2\alpha^{-1} \bigvee_{-1}^{1} h \quad .$$

Since $2\alpha^{-1} < 1$, we find

$$\bigvee_{-1}^{1} f_*^{k-1} h \le (K+2\Delta^{-1})\, ||f_*^{k-1}h||_{L_1} + 2\alpha^{-1} \bigvee_{-1}^{1} f_*^{k-1} h$$

so that

$$\bigvee_{-1}^{1} f_*^{k} h \le (K+2\Delta^{-1})(1-2\alpha^{-1})^{-1}\, ||h||_{L_1} \quad .$$

Thus the set $\{f_*^k h\}_{k=0}^{\infty}$ is relatively compact in L_1. Now the rest of the proof is a standard exercise in analysis. Take, e.g., Dunford-Schwartz as a reference. By Mazur's Theorem, the sequence

$$\left\{ \frac{1}{n} \sum_{k=0}^{n-1} f_*^k h \right\}_{n=0}^{\infty}$$

is relatively compact, too. The set of functions of bounded variation is dense in L_1 and hence we can apply Theorem VIII.5.3 in Dunford-Schwartz [1958] which says that for any $h \in L_1$ we have (i).

Now (ii) is immediate. (iii) is a consequence of our calculations while (iv) and (v) are obvious consequences of the definition of f_*. Thus Lemma 2 is proven and Theorem 1 follows at once by taking $h(x) = 1/2$ for $x \in [-1,1]$.

The measure $h^*d\lambda$ is not necessarily ergodic. An easy example is given by the function f whose graph we draw below.

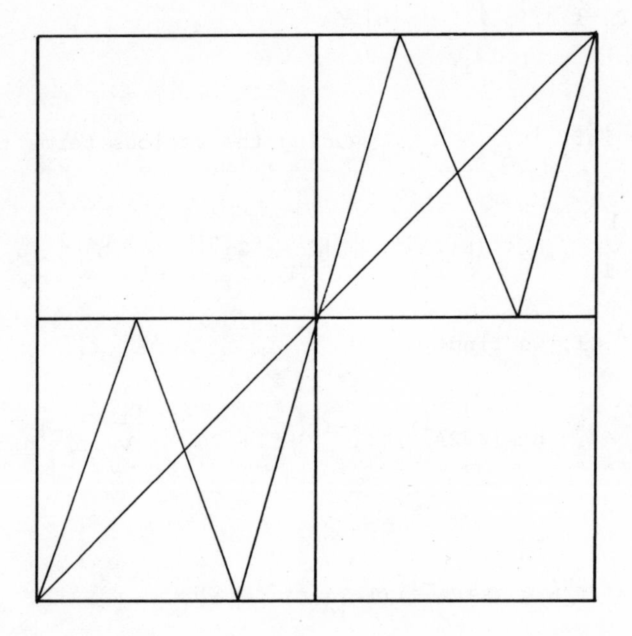

Figure II.23

It clearly does not act transitively on $[-1,1]$, since it leaves $[-1,0]$ and $[0,1]$ invariant. But it satisfies the hypotheses of Theorem 1. We shall see below that for a function with only one extremum (like the unimodal ones) this kind of decomposition is in some situations impossible, so that we shall have a unique invariant measure which is absolutely continuous, and hence it will be ergodic.

We now describe the results of Misiurewicz. Let f denote an S-unimodal function. Then

THEOREM II.8.3. *If f has no stable periodic orbit and and if $0 \notin \{f^n(0): n > 0\}^-$ then f has exactly one absolutely continuous invariant measure. It is ergodic.*

The ergodicity is a direct consequence of the way in which the invariant measure will be constructed. The hypothesis $0 \notin \{f^n(0): n > 0\}^-$ states that the orbit of 0 stays away from zero, and in fact at a finite distance from zero. We do not know about a topological criterion which is necessary and sufficient for this to happen. However, a sufficient condition, going in fact back to the example $f(x) = 1-2x^2$ of Ulam and V. Neumann [1947] has been proposed by several authors, in particular Ruelle [1977]. Namely if 0 is a pre-image of an unstable periodic orbit, then $0 \notin \{f^n(0): n > 0\}^-$. This observation can be transformed into the following

COROLLARY II.8.4. *If f is S-unimodal and $\underline{I}_f(1)$ is eventually periodic, but not periodic, then $f|_{[f(1),1]}$ has an ergodic invariant measure which is absolutely continuous.*

Proof. By Proposition 5.2, f has no stable periodic orbit. By Lemma 3.1, the orbit of 0 converges to a periodic orbit, which must be an unstable one. Thus 0 is not in this periodic orbit, and therefore $|f^n(0) - 0| > \varepsilon$ if $\varepsilon > 0$ is sufficiently small. Thus Theorem 3 applies.

We shall illustrate in Section III.2 a situation where f is (essentially) S-unimodal, satisfies the hypotheses of Theorem 3, but where the orbit of 0 does not converge to an unstable periodic orbit. Misiurewicz [1980] gives similar examples for S-unimodal maps. His theory applies to discontinuous maps with a finite number of discontinuities as well, but we shall (as usual) stick to the S-unimodal maps.

It should be noted that the hypotheses explicitly require that $Df^n(f(0)) \neq 0$ for every $n \geq 0$. It is natural to ask whether some condition as $|Df^n(f(0))| > c^n$ with $c > 1$ would be sufficient to prove existence of an absolutely continuous invariant measure. So far, we know of no such theorem, but it would be very interesting to have one in view of the results of Section III.2.

The structure of Misiurewicz' proof of Theorem 3 is as follows: One first checks that the Lebesgue measure of the set where $|Df^n(x)|$ is small, goes to zero very fast with n. On the complement, one bounds the Perron-Frobenius operator in a way which is reminiscent of the proof of the theorem of Lasota-Yorke. The uniqueness of the invariant measure follows then by "spectral analysis" as in the case of Axiom A diffeomorphisms (cf., e.g. Bowen [1975]).

We proceed through a sequence of lemmas. The first lemma is a direct descendent of Theorem 5.2.3. Let $E_n = \{x : f^j(x) \notin U \text{ for } j = 0,1,\ldots,n-1\}$, where U is a neighborhood of 0.

LEMMA II.8.5. *Let* $f|_{[-1,1]\setminus U} = g$ *have the property that* $|g'| \geq \alpha > 1$. *Let* B *be a subset of* $[-1,1]$ *such that* $f(B) \subset B$ *and* $\beta = \mathrm{dist}(B,U) > 0$. *There is a* $\xi < 1$ *such that for every* $a \in B$

$$\int_{E_n} |x-a|^{-1/2}\, dx \leq (n+1)\xi^n \theta$$

for some universal constant θ.

<u>Proof</u>. Since $f(B) \subset B$, we have $B \subset E_n$ for all n. Consider, instead of $|x-a|^{-1/2}$, the function $\phi(x) = |x-a|^{-1/2} \cdot \chi(\{x > a\})$. We shall show $\int_{E_n} \phi(x)dx \leq (n+1)\xi^n\theta$. A similar argument applies to $|x-a|^{-1/2}\chi(\{x < a\})$, so that the lemma follows from the estimate above. Denote J_k the connected component of $E_k \cap \mathrm{supp}\ \phi$, containing a, for $k = 0,1,2,\ldots$. One of the endpoints of J_k is a, denote the other one by a_k. We have

$$E_n \cap \mathrm{supp}\ \phi = J_n \cup \bigcup_{k=0}^{n-1} H_k ,$$

where $H_k = E_n \cap (J_k \setminus J_{k+1})$. By construction, the endpoints of $f^k(J_{k+1})$ must satisfy $f^k(a) \in B$ and $f^k(a_{k+1}) \in \bar{U}$. We use again the inequality, valid on subintervals K of E_n,

$$\frac{\sup_K |Df^n|}{\inf_K |Df^n|} \leq \delta$$

cf. II.5, Eq. (**) page 115.

Since $\mathrm{dist}(B,U) = \beta > 0$, we find

$$|a-a_{k+1}| \geq \frac{\beta}{\sup_{J_{k+1}} |Df^k|} \geq \frac{\beta}{\sup_{J_k} |Df^k|} \geq \frac{\beta}{\delta \inf_{J_k} |Df^k|} .$$

We now use $f^k(H_k) \subset f^k(E_n) \subset E_{n-k}$ and find

$$\int_{H_k} \phi\ d\lambda \leq \lambda(H_k) \sup_{H_k} \phi \leq \frac{\lambda(f^k(H_k))}{\inf_{H_k} |Df^k|} \ |a-a_{k+1}|^{-1/2}$$

$$\leq \lambda(E_{n-k})\beta^{-1/2}\delta^{1/2} \left(\inf_{J_k} |Df^k|\right)^{-1/2}$$

$$\leq \eta^{n-k}2\beta^{-1/2}\delta^{1/2}(\alpha^{-1/2})^k$$

Finally, we examine the central piece, where we integrate over the singularity: There we have

$$\int_{J_n} \phi \, d\lambda = \int_0^{|a-a_n|} x^{-1/2} \, dx = 2|a-a_n|^{1/2}$$

$$\leq 2 \cdot 2^{1/2} \alpha^{-n/2} ,$$

since $|a-a_n| \leq \alpha^{-n} |f^n(a) - f^n(a_n)| \leq 2\alpha^{-n}$. Choose now $\xi = \max(\eta, \alpha^{-1/2})$ and $\theta = \max(2\beta^{-1/2}\delta^{1/2}, 2^{3/2})$. The result follows.

It is useful to consider a slightly more general situation than the one described in the hypotheses of Theorem 3.

Let A be a finite subset of $J = J(f) = [f(1), 1]$ including the endpoints of J. A is to be thought of as the set of critical points of f plus endpoints of J. Our standard assumptions on f are

(i) f is of class $\mathscr{C}^3$.

(ii) For $x \in J \setminus A$, $f'(x) \neq 0$.

(iii) $Sf(x) \leq 0$.

(iv) If $f^p(x) = x$ then $|Df^p(x)| > 1$.

(v) There is a neighborhood U of A such that for every $a \in A$ and $n \geq 0$, $f^n(a) \in A \cup (J \setminus U)$.

It is clear that the hypotheses of Theorem 3 imply (i)-(v) for f, with $A = \{0, 1, f(1)\}$.

(vi) For every $a \in A$ there is a neighborhood W_a of a and constants $\alpha, \omega > 0$, $u \geq 0$ such that $\alpha|x-a|^u \leq |f'(x)| \leq \omega|x-a|^u$ for $x \in W_a$.

This condition is clearly implied by the stronger condition $f''(0) < 0$ of Theorem 3, we are faced here with the problem of "flat tops." Some condition must be imposed to avoid that the function f is not too near to a function which is constant and equal to 1 near $x = 0$. The absence of stable periodic orbits alone is not sufficient, but probably asking $|Df^p(x)| > g(p,x)$ with g a suitable increasing function of p could be sufficient. It is now useful to make two additional assumptions

(vii) $|f'| > 1$ on $J \setminus U$.

(viii) If $a \in A$ is a periodic point for f, then it is a fixed point for f.

Note that condition (vii) does <u>not</u> follow from the hypotheses of Theorem 3 There we only know $|f'| > \varepsilon > 0$ by the condition of negative Schwarzian derivative. But we shall argue below that some iterate of f satisfies conditions (i)-(viii) (with a different set A). This is the reason for insisting that A may have more than the 3 points $f(1)$, 1, 0.

LEMMA II.8.6. <u>If</u> f <u>satisfies (i)-(vi), then some iterate of</u> f <u>satisfies (i)-(viii)</u>.

<u>Proof</u>. Let $m \geq 1$, and denote $\tilde{f} = f^m$, $\tilde{A} = \cup_{k=0}^{m-1} f^{-k}(A)$. It is easy to see that (i)-(iv) are also satisfied by $\tilde{f}$, $\tilde{A}$ instead of f, A. In (v), take $\tilde{U} = \cup_{k=0}^{m-1} f^{-k}(U)$ instead of U. If (vi) holds for two functions, then a computation shows that it holds for their composition. It is also obvious that if (viii) holds for f then it holds for $\tilde{f}$. So the interesting part is really (vii). It is not trivial, but we have already done the work in Theorem 5.2.2. This completes the proof of Lemma 6.

Denote now by $C_n = \cup_{i=1}^{n} f^i(A)$, $C = \cup_{i=1}^{\infty} f^i(A)$, $B = \bar{C}$. We next give an a priori bound

LEMMA II.8.7. <u>If</u> f <u>satisfies (i)-(iii) then for</u> $x \notin C_n$ <u>one has</u>

$$f_*^n(1)(x) \le \frac{2}{\operatorname{dist}(x,C_n)} .$$

<u>Proof</u>. Let $g = f^n|_L$ where L is a component on which f^n is monotone. We have $Sg \le 0$ and from $g \circ g^{-1} = 1$ it is easy to see that $S(g^{-1}) \ge 0$. It follows from the arguments of Section 4 that $|(g^{-1})'|^{-1/2}$ is concave and hence $|(g^{-1})'|^{1/2}$ is convex and it follows that $g_*(1)$ is also convex, as is easily checked. Thus, if $f^n(L) = (a,b)$, we have either

$$g_*(1)(x) \le \frac{1}{|x-a|} \int_a^x g_*(1)\, d\lambda = \frac{\lambda(g^{-1}(a,x))}{|x-a|} \le \frac{\lambda(L)}{\operatorname{dist}(x,C_n)}$$

or

$$g_*(1)(x) \le \frac{1}{|b-x|} \lambda(g^{-1}(x,b)) \le \frac{\lambda(L)}{\operatorname{dist}(x,C_n)} .$$

Summing over L we obtain the result.

Assume in the sequel for definiteness that condition (vi) is satisfied with $u = 1$ (i.e., $f''(0) < 0$). We analyze the propagation of singularities through g_*.

LEMMA II.8.8. <u>Assume</u> $g:(a,b) \to \mathbb{R}$ <u>is</u> $\mathscr{C}^1$, $g' > 0$, <u>and for</u> $x \in (a,b)$, $\alpha(x-a) \le g'(x) \le \omega(x-a)$, $\alpha, \omega > 0$. <u>Define</u> $\phi: (a,b) \to \mathbb{R}$ <u>by</u> $\phi(x) = (x-a)^{-\zeta}$ <u>with</u> $0 \le \zeta < 1$. <u>Then there is a</u> $\delta > 0$ <u>and</u> $0 \le \xi < 1$ <u>such that</u> $g_*(\phi)(y) \le \delta \cdot (y-g(a))^{-\xi}$ <u>for all</u> $y \in (g(a),g(b))$.

<u>Proof</u>. By definition, $g_*(\phi)(g(x)) = \phi(x)/g'(x) \le \alpha^{-1}(x-a)^{-\zeta-1}$. But $g(x) - g(a) = \int_0^x g'(t)\, dt \le \omega \int_0^{x-a} t\, dt$. Substituting this into the first inequality we find the result with $\xi = (\zeta+1)/2$ and $\delta = \alpha^{-1}(\omega/2)^{\xi}$.

Define now $A_1 = \{a \in A: f(a) = a\}$, $A_2 = A \setminus A_1$. Then we have

LEMMA II.8.9. _For every_ $a \in A_2$ _there are constants_ $\gamma, \delta > 0$ _and_ $\xi, \zeta \in [0,1)$ _such that_

$$f_*^n(1)(x) \le \gamma |x-a|^{-\zeta} \quad \text{for every } n \ge 0 \text{ and } x \in U_a. \tag{8.1}$$

$$(f|_{U_a})_*(f_*^n(1))(y) \le \delta |y-f(a)|^{-\xi} \quad \text{for every } n \ge 0 \text{ and } y \in f(U_a) \tag{8.2}$$

Here U_a is defined as follows. By (v), $B \setminus A$ is disjoint from U. Since $Sf \le 0$ and f' is continuous there are open intervals U_a, $a \in A$ containing $a \in A$ such that $\cup_{a \in A} \bar{U}_a$ is disjoint from $B \setminus A$ and $|f'| \ge \alpha > 1$ on the complement of $\cup_{a \in A} \bar{U}_a$.

Proof of Lemma II.8.9. By (viii) and the definition of A_2, we can order the elements of $A_2 = \{a_1, a_2, \dots, a_j\}$, in such a way that $f(a_i) = a_k$ implies $k > i$. We prove now (8.1) for a_1 then (8.2) for a_1, then (8.1) for a_2 etc., in an inductive manner.

To prove (8.1), note that

$$f_*^n(1) = (f|_G)_*(f_*^{n-1}(1)) + (f|_{J \setminus G})_*(f_*^{n-1}(1)),$$

and we choose

$$G = \bigcup_{j \in T} U_{a_j}$$

where

$$T = \{j : f(a_j) = a_i\}.$$

Then we obtain an estimation of the first summand from (8.2) for $j \in T$, and the second one from Lemma 7 and the fact that by the definition of U_a, $\text{dist}(B \setminus A, \bigcup_{a \in A} U_a) > 0$. We get a finite sum of expressions of the form $\gamma|x-a_i|^{-\zeta}$ (notice that a constant is also of this form for $\zeta = 0$), and the sum is not greater than some function of the same form. Thus we obtain (8.1), but in general only for x from some smaller semi-neighborhood of a_i than U_{a_i}. Using once more Lemma 7 we obtain an estimation by a constant on the rest of U_{a_j}. Again, the sum of a function of a form $\gamma|x-a_i|^{-\zeta}$ and a constant is not greater than some function of the form $\gamma|x-a_i|^{-\zeta}$.

(8.2) for a_i follows from (8.1) for a_i and Lemma 8. The next lemma will be useful for bounding $f_m^*(1)$ on those x whose orbit will avoid U for m iterations.

LEMMA II.8.10. *Let* $H \subset J$ *and* $H_k = \{x: f^i(x) \in H$ *for* $i = 0,\dots,k-1\}$. *Then for every* s *and* m *one has*

$$\int_{H_s} f_*^m(1)\, d\lambda \le \sum_{k=s}^{\infty} \int_{H_k} \sup_{n \ge 0} \left(f|_{J \setminus H}\right)_* (f_*^n(1))\, d\lambda + \lambda(H_{s+m}).$$

Proof. Let $G_k = f^{-1}(H_k) \setminus H$. We have $f^{-1}(H_k) = G_k \cup H_{k+1}$, and by induction we get

$$f^{-m}(H_s) = \bigcup_{k=s}^{s+m-1} f^{-s-m+1+k}(G_k) \cup H_{s+m} .$$

Hence

$$\int_{H_s} f_*^m(1)\, d\lambda = \lambda(f^{-m}(H_s)) \le \sum_{k=s}^{s+m-1} \lambda(f^{-s-m+1+k}(G_k)) + \lambda(H_{s+m})$$

$$= \sum_{k=s}^{s+m-1} \int_{f(G_k)} \left(f|_{G_k}\right)_* (f_*^{s+m-1-k}(1))\, d\lambda + \lambda(H_{s+m})$$

$$\leq \sum_{k=s}^{\infty} \int_{H_k} \sup_{n\geq 0} (f|_{J\setminus H})_*(f_*^m(1))\, d\lambda + \lambda(H_{s+m}) \; .$$

The next lemma deals with $a \in A_1$. This is the case of repelling fixed points.

LEMMA II.8.11. <u>For every</u> $a \in A_1$ <u>and</u> $\varepsilon > 0$ <u>there exists a neighborhood</u> W <u>of</u> a <u>such that</u>

$$\int_{W\cap U_a} f_*^n(1)\, d\lambda < \varepsilon \qquad \underline{\text{for every}} \quad n \geq 0.$$

<u>Proof</u>. Let $a \in A_1$. Since $f(a) = a$ is an unstable fixed point we have $|f(x)-a| \geq \beta|x-a|$ for some $\beta > 1$, and for all $x \in U_a$. Denote $V_k = \{x\colon f^i(x) \in U_a$ for $i = 0,\dots,k-1\}$. By the above inequality we have

$$\lambda(V_k) \leq 2/\beta^k \quad , \qquad (*)$$

and thus it suffices to prove

$$\sup_{m\geq 0} \int_{V_s} f_*^m(1)\, d\lambda \to 0 \qquad \text{as} \qquad s \to \infty \quad .$$

We decompose

$$\left(f\Big|_{I\setminus U_a}\right)_* (f_*^n(1)) = \left(f\Big|_G\right)_* (f_*^n(1)) + \left(f\Big|_{I\setminus(G\cup U_a)}\right)_* (f_*^n(1)),$$

where $G = \cup_{b\in R} U_b$, $R = \{b \in A\setminus\{a\}\colon f(b) = a\}$. The same arguments as in the proof of Lemma 9 show that

$$\sup_{n\geq 0} \left(f\Big|_{J\setminus U_a}\right)_* (f_*^n(1))(y) \leq \delta|y-a|^{-\xi}$$

for some constants $\delta > 0$, $0 \leq \xi < 1$ (notice that $R \subset A_2$, so we can use (8.2)). In view of Lemma 10 (for $H = U_a$; then $H_k = V_k$) and (*) we get

$$\sup_{m\geq 0} \int_{V_s} f_*^m(1)\, d\lambda \leq \sum_{k=s}^{\infty} \int_0^{\lambda(J)/\beta^k} t^{-\xi}\, dt + \frac{\lambda(J)}{\beta^{s+m}}$$

$$\leq \delta \sum_{k=s}^{\infty} \frac{1}{1-\xi} \left(\frac{\lambda(J)}{\beta^k}\right)^{1-\xi} + \frac{\lambda(J)}{\beta^s}$$

$$\leq \frac{\delta}{1-\xi}\, \lambda(J)^{1-\xi} \sum_{k=s}^{\infty} (\beta^{\xi-1})^k + \frac{\lambda(J)}{\beta^s} \to 0$$

as $s \to \infty$. This proves the lemma.

Next we show the bound analogous to Lemma 11 for $a \in B \setminus A$.

LEMMA II.8.12. <u>For every</u> $\varepsilon > 0$ <u>there exists a neighborhood</u> W <u>of</u> $B \setminus A$ <u>such that</u>

$$\int_W f_*^n(1)\, d\lambda < \varepsilon \qquad \text{for all} \quad n \geq 0.$$

<u>Proof</u>. Define $\bigcup_{a\in A} U_a = U$ and $B' = B \setminus A$. Let $E_n = \{x \in J\colon f^k(x) \notin U$ for $k = 0,\dots,n-1\}$ and let $\beta = \mathrm{dist}(B',U)$. We have $\beta > 0$. U and B' satisfy the hypotheses of Lemma 5, and E_n is a neighborhood of $B \setminus A$. Hence it suffices to prove

$$\sup_{m\geq 0} \int_{E_s} f_*^m(1)\, d\lambda \to 0 \qquad \text{as} \quad s \to \infty \quad . \qquad (*)$$

By Proposition II.5.2.2 and Lemma 9 there exist constants $\eta,\xi \in (0,1)$ and $\theta > 0$ such that $\lambda(E_s) \leq \eta^s \lambda(J)$ for all $s \geq 0$ and

$$\int_{E_s} \sup_{m\geq 0} \left(f \,\Big|\, \bigcup_{a\in A} U_a\right)_* (f_*^m(1))\, d\lambda \leq (s+1)\theta\xi^s \quad \text{for all} \quad s \geq 0.$$

Hence, by Lemma 10 (for $H = J \setminus \bigcup_{a \in A} U_a$, $H_k = E_k$) we obtain

$$\sup_{m \geq 0} \int_{E_s} f_*^m(1)\, d\lambda \leq \sum_{k=s}^{\infty} (s+1)\theta\xi^s + \lambda(J)\eta^s \to 0$$

as $s \to \infty$. This proves (*) and hence Lemma 12.

We finally prove the existence of an absolutely continuous invariant measure in

LEMMA II.8.13. _If_ f _satisfies (i)-(vi) then for every_ $\varepsilon > 0$ _there exists_ $\delta > 0$ _such that if_ $G \subset J$ _and_ $\lambda(G) < \delta$ _then_ $\int_G f_*^n(1)\, d\lambda < \varepsilon$ _for all_ n.

Proof. Suppose first that f satisfies (i)-(viii). Since a function of the form $\phi(x) = \gamma|x-a|^{-\zeta}$ $(0 \leq \zeta < 1)$ is integrable at a, it follows from Lemma 9 that the statement of Lemma 11 holds also for all $a \in A_2$. This and Lemmata 11 and 12 imply that for every $\varepsilon > 0$ there exists a neighborhood W of B such that

$$\int_W f_*^n(1)\, d\lambda < \varepsilon \qquad \text{for all } n.$$

Now suppose that f satisfies (i)-(vi). By Lemma 6, there exists $m \geq 1$ such that f^m satisfies (i)-(viii). In order to distinguish the sets A and B defined for different iterates of f we shall use symbols $A(f^k)$ and $B(f^k)$ for those sets defined for f^k. There exists $q \geq 1$ such that the set $J \setminus A(f^k)$ consists of at most q components for $k = 0,1,\ldots,m-1$.

Fix $\varepsilon > 0$. Since f^m satisfies (i)-(viii), there exists an open neighborhood W of $B(f^m)$ such that

$$(1) \qquad \int_W f_*^{mn}(1)\, d\lambda < \frac{\varepsilon}{3} \qquad \text{for all } n \geq 0 .$$

Take

$$(2) \qquad \eta = \frac{\varepsilon \operatorname{dist}(B(f^m), J\setminus W)}{3\lambda(J)}$$

Clearly $\eta > 0$. There exist neighborhoods U_k of $A(f^k)$, $k = 0,1,\dots,m-1$ such that

$$(3) \qquad \lambda(U_k) \le \eta \quad .$$

Take

$$(4) \qquad \delta = \frac{\eta}{q} \min_{0\le k\le m-1} \inf_{J\setminus U_k} |Df^k| \quad .$$

By (ii), also $\delta > 0$.

Suppose now that $G \subset J$ and $\lambda(G) < \delta$. Fix $k \in [0,m-1]$. We want to estimate $\int_G f_*^{mn+k}(1)\, d\lambda$. Let $J_1,\dots,J_p$ be the components of $J \setminus A(f^k)$. By the definition of q, we have $p \le q$. We have

$$(5) \qquad \int_G f_*^{mn+k}(1)\, d\lambda = \int_{f^{-k}(G)} f_*^{mn}(1)\, d\lambda$$

$$\le \int_W f_*^{mn}(1)\, d\lambda + \int_{U_k\setminus W} f_*^{mn}(1)\, d\lambda$$

$$+ \sum_{i=1}^{p} \int_{f^{-k}(G)\cap J_i\setminus(U_k\cup W)} f_*^{mn}(1)\, d\lambda.$$

By (2), (3) and Lemma 7 we have

$$(6) \qquad \int_{U_k\setminus W} f_*^{mn}(1)\, d\lambda \le \lambda(U_k)\cdot \frac{\lambda(J)}{\operatorname{dist}(J\setminus W, B(f^m))} \le \frac{\varepsilon}{3} \quad .$$

By (2), (3), (4) and Lemma 7 we have

$$(7) \qquad \sum_{i=1}^{p} \int_{f^{-k}(G)\cap J_i \smallsetminus (U_k \cup W)} f_*^{mn}(1)\, d\lambda$$

$$\leq p\cdot\lambda(f^{-k}(G)\cap J_i \smallsetminus U_k)\cdot \frac{\lambda(J)}{\operatorname{dist}(J\smallsetminus W, B(f^m))}$$

$$\leq q\cdot \frac{\lambda(G)}{\inf\limits_{J_i\smallsetminus U_k} |(f^k)'|} \cdot \frac{\lambda(I)}{\operatorname{dist}(J\smallsetminus W, B(f^m))}$$

$$< \frac{q\cdot\delta}{\inf\limits_{I\smallsetminus U_k} |(f^k)'|} \cdot \frac{\lambda(J)}{\operatorname{dist}(J\smallsetminus W, B(f^m))} \leq \frac{\varepsilon}{3} \quad .$$

Now from (5), (1), (6) and (7) we obtain $\int_G f_*^{mn+k}(1)\, d\lambda < \varepsilon$.

From Lemma 13, it follows that any weak-* limit of a subsequence of $\{1/n\ \Sigma_{k=0}^{n-1} f_*^k(\lambda)\}_{n=1}^{\infty}$ is an invariant measure, absolutely continuous with respect to Lebesgue measure. A very important question is now whether this measure is unique and ergodic. Under the hypotheses of Theorem 3, Misiurewicz proves that this is indeed the case. We describe only the general steps of his proof and refer to the original paper [1980] for details.

First it is easy to see from the previous estimates that the set $B = (\cup_{n=0}^{\infty} f^n(A))^-$ is not very large. In fact Lemma 12 implies $\lambda(B) = 0$. This illustrates simultaneously the beauty and the weakness of the idea that $\{f^n(0)\}_{n>0}^- \not\ni 0$. In fact it seems that it is much more "probable" for a map to satisfy $\{f^n(0)\}_{n>0}^- \ni 0$ as we shall see in Section III.2. However, no proof of existence of an absolutely continuous invariant measure is known in this general case.

The next step in the argument is to study the convexity of $f_*^n(1)$ on the intervals of $J \smallsetminus B$, by an extension of the ideas used in the proof of Lemma 7.

Now one has the necessary tools to study the spectral decomposition as in the case of Axiom-A diffeomorphisms. Denote by $\mathcal{J}$ the set of all components of $J \setminus B$. Since $f^n(B) \subset B$ for all n, we see that if $K, L \in \mathcal{J}$ and $f^n(K) \cap L \neq \emptyset$ then $f^n(K) \supset L$. Hence the dynamical system is something like a Markoff chain with a countable (infinite) number of states.

We shall say $K \sim L$ iff there is an $n \geq 0$ such that $f^n(K) \supset L$, and we define $\mathcal{H}$ as those $K \in \mathcal{J}$ such that for all $L \in \mathcal{J}$ one has: if $K \sim L$ then $L \sim K$. Then one shows that $\sim$ is an equivalence relation in $\mathcal{H}$ and has only one equivalence class in $\mathcal{H}$.

One also says $K \approx L$ iff there is an $n \geq 0$ such that $f^n(K) \supset K \cup L$. We denote by $\mathcal{K} = \{\overline{\cup \mathcal{G}} : \mathcal{G}$ is an equivalence class of $\approx$ in $\mathcal{H}\}$. Then one proves $\mathcal{K}$ is finite and f maps elements of $\mathcal{K}$ onto elements of $\mathcal{K}$. Furthermore, for every K in $\mathcal{K}$ there is an m for which $f^m(K) = K$, and for every $J \in \mathcal{J}$, $K \in \mathcal{K}$ with $J \subset K$ there is an m such that $f^m(J) \supset K$. From this one sees that each $K \in \mathcal{K}$ is an interval.

It follows that for every $K \in \mathcal{K}$ there is a probability measure μ_K, absolutely continuous and f^k-invariant where $f^k(K) = K$. This measure is the unique abs.cont. f^k-invariant measure. The system $(K, f^k|_K, \mu_K)$ is exact (see Walters [1975(2)] for the definition). Now one obtains easily that in the case of unimodal maps there is a unique invariant measure which is absolutely continuous. Hence it is ergodic.

Entropy

There are interesting invariants for maps on the interval, in fact for maps in general, which we have not described so far, especially in view of the fact that the corresponding results are yet too weak to lead to observable consequences. To do justice to the interesting mathematical work, we list here the main definitions and results, and some references.

There are two main notions of entropy used in the subject, and we now give their definitions.

The topological entropy of f (when considering unimodal maps f) is defined as

$$h(f) = \lim_{n\to\infty} \frac{1}{n} \log N_n$$

where N_n is the number of laps (i.e., monotone pieces) of f^n. There are alternate definitions and theorems relating these definitions (Misiurewicz [1976], Adler-Konheim-McAndrew [1965], Misiurewicz-Slenk [1977, 1980]).

Several papers deal with the problem of establishing lower bounds for $h(f)$. (Bowen-Franks [1976], Jonker-Rand [1980], Misiurewicz-Slenk [1977, 1980], Milnor-Thurston [1977], Block-Guckenheimer-Misiurewicz-Young [1979]). The best bound seems to be

THEOREM II.8.14. If f has a periodic point of period $p\cdot 2^m$ where p is odd and $p > 1$ then $h(f) \geq 2^{-m} \log \lambda_p$ where λ_p is the largest (real) root of $x^p - 2x^{p-2} - 1$.

There are also results about the dependence of $h(f)$ on f. If f' and f'' do not vanish simultaneously, then $h(f)$ is continuous at $f \in \mathscr{C}^2([-1,1] \to [-1,1])$.

The measure-theoretic entropy $h(\rho,f)$ is defined for every probability measure ρ which is invariant under f. If $\alpha = \{A_i\}_{i\in I}$ is a partition of $(-1,1)$ into measurable subsets, then $H(\rho,\alpha)$ (the entropy of the partition) is defined by

$$H(\rho,\alpha) = \sum_{i\in I} \rho(A_i)\,|\log \rho(A_i)| \quad .$$

If α is a partition then $f^{-1}\alpha$ denotes the partition defined by the sets $\{f^{-1}A_i\}_{i\in I}$. Then one defines

$$h(\rho,\alpha,f) = \lim_{n\to\infty} \frac{1}{n} H(\rho,\alpha \vee f^{-1}\alpha \vee \ldots \vee f^{-n+1}\alpha)$$

where $\vee$ is common refinement.

The <u>measure-theoretic entropy</u> is then defined by

$$h(\rho,f) = \sup_{\alpha} h(\rho,\alpha,f)$$

where $\sup_\alpha$ is over all partitions of $[-1,1]$ for which $H(\rho,\alpha) < \infty$.

LEMMA II.8.15. *If ξ is a finite partition and ρ is ergodic then*

$$h(\rho,f) = \lim_{n\to\infty} -\frac{1}{n} \log \rho(\xi_n(x))$$

for ρ-a.e. x, *where*

$$\xi_n = \xi \vee f^{-1}\xi \vee \ldots \vee f^{-n+1}\xi \quad ,$$

and $\xi(x)$ *is the element of* ξ *containing* x.

References: Sinai [1976], Adler-Konheim-McAndrew [1965], Bowen [1979].

An interesting question is now the study of the relation between the topological and measure-theoretic entropy.

THEOREM II.8.16. _One has_ $h(\rho,f) \leq h(f)$. There are many references for this, but the most complete account is probably Ruelle's book [1979(2)].

Consider still the situation where f preserves a measure ρ. _Then_

$$\lambda(x) = \lim_{n\to\infty} \frac{1}{n} \log|Df^n(x)|$$

exists ρ-almost-everywhere. Oseledec [1978], Ragunathan [1980], Ruelle [1978(1)]. In fact these authors prove much more general results. The numbers $\lambda(x)$ are called the _characteristic exponents_ or _Liapunov exponents_.

Define $\lambda_+(x) = \max(\lambda(x),0)$. _Then the entropy satisfies_

$$h(\rho,f) \leq \int \lambda_+(x)\, d\rho(x) \quad .$$

Note that if ρ is ergodic, the $\lambda_+(x) = \text{const} = \lambda_+$ ρ-almost-everywhere, and the inequality provides a lower bound for the Liapunov characteristic exponent. However, still too many relations between the quantities defined above seem ill-understood, and so we have refrained from giving any further interpretation. General references: Ruelle [1979(2), 1978(1), 1979(1)], Bowen-Ruelle [1975].

Remarks and Bibliography. Up to a recent past, ergodic properties have been shown almost exclusively for maps f with $|f'| > 1$, and there is a vast literature on the subject. The easiest cases are those where the invariant measure can be explicitly calculated: Ulam-v. Neumann [1947], for the case of $x \to 1-2x^2$; the example $x \to$ fractional part of $1/x$ is attributed to Gauss (see Renyi [1957] as a general reference). The next stage of generalization is that of piecewise linear maps. This has been extensively studied by Parry [1964]. When f is a piecewise $\mathscr{C}^2$ function and $|f'| > 2$, then by the theorem of Lasota-Yorke [1973] one has an absolutely continuous invariant measure. Kowalski [1976] has shown how this measure is decomposable into ergodic components. This circle of ideas has been generalized further by Walters [1975(1)]. A new class of functions with critical points became treatable through the ideas of Ruelle [1977], Bowen [1979], Pianigiani [1979]. Here one considers maps whose critical point falls on an unstable periodic orbit. This idea was taken up and brought to perfection by Misiurewicz [1980]. Another generalization still partly unpublished is due to Jakobson [1978], [1979]. It uses extensively the idea of return maps (also called induced maps). The author considers maps such that the induced map has very large derivative wherever it is defined.

PART III. PROPERTIES OF ONE-PARAMETER FAMILIES OF MAPS

III.1 ONE-PARAMETER FAMILIES OF MAPS

The purpose of this section is to show certain combinatorial regularities for one parameter families of maps. These were discovered and most systematically exploited by Metropolis, Stein, Stein [1973].

Throughout this section $\mu \to f_\mu$ denotes a curve in the space of $\mathscr{C}^1$-unimodal maps, continuous in the $\mathscr{C}^1$ topology. The precise definition is

$$\mu \mapsto f_\mu$$

is a map from $[0,1]$ to the $\mathscr{C}^1$-unimodal maps such that

$$\sup_{x\in[-1,1]} |f_\mu(x) - f_{\mu_0}(x)| + |f'_\mu(x) - f'_{\mu_0}(x)| \to 0$$

when $\mu \to \mu_0$. Let us write $\underline{K}(f_\mu)$ for the itinerary $\underline{I}_{f_\mu}(1)$. The main result of this section is the

THEOREM III.1.1. <u>Let</u> $\underline{A}$ <u>be a maximal sequence satisfying</u> $\underline{K}(f_0) < \underline{A} < \underline{K}(f_1)$. <u>Then there is a</u> $\mu \in (0,1)$ <u>such that</u> $\underline{K}(f_\mu) = \underline{A}$.

The analogy between this theorem and Theorem II.3.8 is obvious. This is probably the reason for the confusion about the occurrence of <u>stable</u> periodic orbits in Šarkovskii's Theorem. In fact, for one parameter families, we see that if $\underline{A}$ is <u>finite</u> and $\underline{K}(f_\mu) = \underline{A}$ then f_μ will have a <u>superstable</u> period of type $\underline{A}$. So in particular the theorem tells us that if a one-parameter family of maps has a superstable period of type $\underline{K}(f_0)$, (i.e., $\underline{K}(f_0)$ is finite) for $\mu = 0$ and a superstable period of type $\underline{K}(f_1)$ for $\mu = 1$ then <u>all</u>

superstable periods with $\underline{I}(1)$ between $\underline{K}(f_0)$ and $\underline{K}(f_1)$ will occur. It should be stressed, however, that the theorem does not claim that every $\underline{A}$ occurs only once, and it is possible that the family "oscillates" several times between $\underline{K}(f_0)$ and $\underline{K}(f_1)$ when μ is varied from 0 to 1. We shall see only in Section III.3 a very notable exception in a "neighborhood" (in the sense of the ordering) of the itinerary "$R*^{\infty}*RC$".

We shall call a family full if $f_0(1) = 0$ and $f_1(1) = f_1(-1) = -1$. Then we have

PROPOSITION III.1.2. In a full family of $\mathscr{C}^1$-unimodal maps every maximal itinerary of the form R... occurs as the $\underline{K}(f_\mu)$ for some $\mu \in [0,1]$.

Proof of Theorem III.1.1. The proof is very similar to that of Theorem II.3.8. We first claim that if $\underline{A} \neq (\underline{B}R)^{\infty}$ or $(\underline{B}L)^{\infty}$ then

$$M_{\underline{A}} \equiv \{f: f \text{ is } \mathscr{C}^1\text{-unimodal and } \underline{K}(f) < \underline{A}\}$$
$$P_{\underline{A}} \equiv \{f: f \text{ is } \mathscr{C}^1\text{-unimodal and } \underline{K}(f) > \underline{A}\} \quad (*)$$

are open in the $\mathscr{C}^1$-topology of $\mathscr{C}^1$-unimodal maps.

Proof. Let $\underline{D} = \underline{K}(g) < \underline{A}$ (the case of $\underline{D} > \underline{A}$ is analogous and is left to the reader). We need to find a neighborhood of g contained in $M_{\underline{A}}$. If i is the first index for which $D_i \neq A_i$ and if $D_i \neq C$, then $\{f \in M_{\underline{A}}: K_j(f) = D_j$ for $j = 1,\ldots,i\}$ is open, contains g and is contained in $M_{\underline{A}}$. The only remaining possibility is $\underline{D} = \underline{B}C$, $\underline{A} = \underline{B}\underline{X}$ and $\underline{X} \neq C$. This is dealt with by an exact analog of Lemma II.3.9.

LEMMA III.1.3. Let g be $\mathscr{C}^1$-unimodal and superstable, with $\underline{I}_g(1) = \underline{B}C$. There is a $\mathscr{C}^1$-neighborhood U of g (in the space of $\mathscr{C}^1$-unimodal maps) such that for all $f \in U$,

$\underline{K}(f)$ __is one of the three sequences__ $\underline{B}C$, $(\underline{B}R)^{\infty}$ __or__ $(\underline{B}L)^{\infty}$.

__Proof__. There exists a $\mathscr{C}^1$-neighborhood U of g such that

1. If $f \in U$ and $|x| < \varepsilon$ then $\underline{I}_f(x) = \underline{B}\ldots$.

2. If $|\underline{B}C| = n$, then $|Df^n(x)| < 1/2$ for $f \in U$ and $|x| < \varepsilon$. (Because $Df^n(0) = 0$.)

3. If $f \in U$ then $|f^n(x)| < \varepsilon/2$, when $|x| < \varepsilon$.

(First choose U, ε so that 1, 2 hold; then if necessary make U smaller so that 3 holds.) Only 2 requires that f be near g in the $\mathscr{C}^1$ topology rather than the weaker $\mathscr{C}^0$ topology. We now show that U is the adequate neighborhood.

If $f \in U$ and $f^n(0) = 0$, then $\underline{K}(f) = \underline{B}C$. So let now $f \in U$ and $f^n(0) = a > 0$. Let $x \in (0,2a)$. Then, since $2a < \varepsilon$ by 3, we have $|f^n(x) - f^n(0)| \le |x|/2$ by 2. Thus $f^n(x) \ge f^n(0) - 1/2|x| > a - 1/2\ 2a = 0$ and $f^n(x) \le f^n(0) + 1/2|x| < a + 1/2\ 2a = 2a$ so that $f^n(0,2a) \subset (0,2a)$. Thus $\underline{K}(f) = (\underline{B}R)^{\infty}$. Similarly if $a < 0$ then $\underline{K}(f) = (\underline{B}L)^{\infty}$. Q.E.D.

We complete now the proof of Theorem 1. By (*) if $\underline{A}$ is not $(\underline{B}L)^{\infty}$ or $(\underline{B}R)^{\infty}$, then

$$M^{\perp}_{\underline{A}} \equiv \{\mu \in [0,1]: \underline{K}(f_{\mu}) \le \underline{A}\}$$

and

$$P^{\perp}_{\underline{A}} \equiv \{\mu \in [0,1]: \underline{K}(f_{\mu}) \ge \underline{A}\}$$

are closed. Since $\underline{K}(f_0) < \underline{A} < \underline{K}(f_1)$ they are both nonempty, their union is $[0,1]$. Since $[0,1]$ is connected, $M^{\perp}_{\underline{A}} \cap P^{\perp}_{\underline{A}}$ cannot be empty and thus $\underline{K}(f_{\mu}) = \underline{A}$ for some $\mu \in [0,1]$. If $\underline{A}$ is $(\underline{B}L)^{\infty}$, we argue as follows. By the argument just given $\{\mu \in [0,1],\ \underline{K}(f_{\mu}) = \underline{B}C\}$ is closed and not empty. Let $\mu_1 = \inf\{\mu \in [0,1]: \underline{K}(f_{\mu}) = \underline{B}C\}$. For μ slightly less than μ_1,

$\underline{K}(f_\mu) \neq \underline{B}C$, and in fact $\underline{K}(f_\mu) < \underline{B}C$ (otherwise μ_1 is not minimal). Since μ is near μ_1, we find $\underline{K}(f_\mu) = (\underline{B}L)^\infty$ or $(\underline{B}R)^\infty$ whichever of the two is less than $\underline{B}C$ (depending on the parity of $\underline{B}$). Similarly we can argue with the supremum to find the missing sequence $(\underline{B}R)^\infty$ or $(\underline{B}L)^\infty$, respectively. This completes the proof of Theorem 1.

Proof of Proposition III.1.2. Since $\{f_\mu\}$ is full, $\underline{K}(f_0) = RC$, $\underline{K}(f_1) = RL^\infty$. The assertion is now immediate from Theorem 1.

Note that the continuity in the $\mathscr{C}^1$ topology is essential for the validity of the theorem. As an example, we discuss the case of the piecewise linear maps $g_\mu(x) = 1 - \mu|x|$.

We need some definitions. Let $\underline{A}C$ be maximal. If $\underline{A}C > RLR^\infty$, we say $\underline{A}C$ is primary if it cannot be written as $\underline{B}*\underline{D}C$ with $\underline{B} \neq \emptyset$, $\underline{D} \neq \emptyset$. If $\underline{A}C \leq RLR^\infty$ is not equal to $R*^m*RC$ for some $m \geq 0$, then $\underline{A}C$ can, by Theorem II.2.7, be written as $\underline{A}C = R*^n*\underline{B}C$, for some (unique) $n > 0$ and $\underline{B}C > RLR^\infty$, since $RLR^\infty = R*RL^\infty$. Then $\underline{A}C \leq RLR^\infty$ will be called primary if $\underline{B}C$ is primary. If $\underline{A}C = R*^m*RC$, $m > 0$ it is not primary, RC is primary.

LEMMA III.1.4. (Derrida-Gervois-Pomeau [1979].) If $\underline{A}C$ is not primary, then there is no $\mu \in (0,2]$ for which $\underline{K}(g_\mu) = \underline{A}C$. If $\underline{A}C = R*^m*RC$, there is no $\mu \in (0,2]$ for which $\underline{K}(g_\mu) = \underline{A}C$ if $m > 0$, and exactly one if $m = 0$ (and then $\mu = 1$).

We shall present most of the proof, since it is interesting in its own right, but we shall refer to the original proof for one tedious calculation. To analyze the situation described in Lemma 4, we develop a calculus of polynomials associated to itineraries not containing C. Define $\varepsilon(L) = +1$, $\varepsilon(R) = -1$. Assume $\underline{K}(g_\mu) = B_0B_1B_2\dots$, where $g_\mu(x) = 1 - \mu|x|$, with $\mu \in (1,2]$. Then

$$g_\mu(1) = 1-\mu = 1+\varepsilon(B_0)\mu$$

$$g_\mu^2(1) = 1+\mu\varepsilon(B_1)(1+\varepsilon(B_0)\mu)$$

and inductively

$$g_\mu^k(1) = 1 + \sum_{j=1}^{k} \mu^j \prod_{\ell=1}^{j} \varepsilon(B_{k-\ell}) \quad .$$

To a <u>finite</u> sequence $\underline{B}$ of L's and R's we associate the <u>characteristic polynomial</u>

$$\mathscr{P}_{\underline{B}C}(\mu) = 1 + \sum_{j=1}^{|\underline{B}|} \mu^j \prod_{\ell=1}^{j} \varepsilon(B_{|\underline{B}|-\ell}) \quad .$$

This polynomial is closely related to the <u>kneading determinant</u> of Milnor-Thurston. The condition that $\underline{K}(g_\mu) = \underline{B}C$ implies that μ is a real solution >0 of $\mathscr{P}_{\underline{B}C}(\mu) = 0$.

The calculus of the *-product is now reflected through

LEMMA III.1.5.

$$\mathscr{P}_{\underline{A}*\underline{B}C}(\mu) = \mathscr{P}_{\underline{A}C}(\mu)\cdot\mathscr{P}_{\underline{B}C}(\mu^n)$$

<u>where</u>

$$n = |\underline{A}C| \quad .$$

<u>Proof</u>. The proof is purely calculatory. Define

$$\mathscr{Q}_{\underline{B}C}(\tfrac{1}{\mu}) = \mu^{-|\underline{B}|}\varepsilon(B_0)\ldots\varepsilon(B_{|\underline{B}|-1})\,\mathscr{P}_{\underline{B}C}(\mu) \quad .$$

Then, if $\underline{A}$ is even, and $|\underline{A}C| = n$, we find

$$\mathcal{Q}_{\underline{A}*\underline{B}C}(z) = 1 + z\varepsilon(A_0) + \ldots + z^{n-1}\varepsilon(A_0)\varepsilon(A_1)\ldots\varepsilon(A_{n-2})$$

$$+ z^n\varepsilon(B_0) + z^{n+1}\varepsilon(B_0)\varepsilon(A_0) + \ldots \quad ,$$

since

$$\varepsilon(A_0)\ldots\varepsilon(A_{n-2}) = 1.$$

Thus we see that

$$\mathcal{Q}_{\underline{A}*\underline{B}C}(z) = \mathcal{Q}_{\underline{A}C}(z)\,\mathcal{Q}_{\underline{B}C}(z^n) \tag{1}$$

in this case. When $\underline{A}$ is odd, then we have $\underline{A}*\underline{B}C = \underline{A}\check{B}_0\underline{A}\check{B}_1\ldots$ and the equality (1) follows from $\varepsilon(B_i) = -\varepsilon(\check{B}_i)$. Resubstituting the definition of $\mathcal{Q}$ as a function of $\mathcal{P}$ the assertion of the lemma follows.

Suppose that for some value of μ, $\underline{K}(g_\mu) = \underline{B}C$. Then it follows from the definition of $\mathcal{P}_{\underline{B}C}$ that for this value of μ, $\mathcal{P}_{\underline{B}C}(\mu) = 0$. But not every zero of this equation is a possible choice of the parameter, because we must insure that the whole trajectory is correctly reproduced. Let us write for this purpose $\underline{B}C = RL^{Q(\underline{B})}R\ldots$ or $\underline{B}C = RL^{Q(\underline{B})}C$. Thus if $\underline{K}(g_\mu) = \underline{B}C$, then μ must satisfy in particular the inequalities $g_\mu^{Q(\underline{B})}(1) < 0$ and $g_\mu^{Q(\underline{B})+1}(1) \geq 0$. Expressed as polynomial inequalities, this becomes

$$\sum_{j=0}^{Q(\underline{B})-1} \mu^j - \mu^{Q(\underline{B})} < 0 \tag{2}$$

and

$$\sum_{j=0}^{Q(\underline{B})} \mu^j - \mu^{Q(\underline{B})+1} \geq 0 \quad , \tag{3}$$

as is easily seen from the definitions. Let now $\mu^{(Q)}$ be the largest real solution of

$$\frac{1-\mu^Q}{1-\mu} - \mu^Q = 0 \quad .$$

It is easy to see that there is exactly one solution >1 when $Q > 1$. Therefore our original μ must satisfy $\mu^{(Q(\underline{B}))} < \mu \leq \mu^{(Q(\underline{B})+1)}$, since $\mu^{(Q+1)} > \mu^{(Q)}$. Defining $\mu^{(1)} = 1$, we also cover the case $Q = 1$.

Next we consider a general $\underline{B}C > RLR^{\infty}$, which can always be written in the form

$$\underline{B}C = RL^{n_1}RL^{n_2}R\ldots RL^{n_p}C \quad ,$$

with $n_i \geq 0$, $n_1 > 1$. The maximality of $\underline{B}C$ forces in particular the inequalities $n_i \leq n_1$, $i = 3,4,\ldots,p$, and $n_2 < n_1$, as can be easily checked. Using a more tedious variant of the argument given above (an induction over p, see Appendix B in Derrida-Gervois-Pomeau [1979]), one can show: Let $\underline{B}C > RLR^{\infty}$.

If $n_1 \geq 2$, then $\mathscr{P}_{\underline{BC}}(\mu) = 0$ has exactly one root in $\{\mu : \mu > 1\}$, and this root $\mu_{\underline{BC}}$ satisfies

$$\mu^{(n_1)} < \mu_{\underline{BC}} \leq \mu^{(n_1+1)} \quad .$$

If $n_1 = 1$, then $\mathscr{P}_{\underline{BC}}(\mu) = 0$ has exactly one root in $\{\mu : \mu > 1\}$, and this root $\mu_{\underline{BC}}$ satisfies $\sqrt{2} < \mu_{\underline{BC}} < \mu^{(2)}$. The above results imply in particular that the roots are larger than $\sqrt{2}$. Note that $\sqrt{2}$ is characterized by $\underline{K}(g_{\sqrt{2}}) = RLR^{\infty}$.

We can now complete the proof of Lemma 4. By Theorem II.2.7, if $RC < \underline{B}C \leq RLR^{\infty} = R{*}RL^{\infty}$, then $\underline{B}C = R{*}\underline{B}_1C$ and if $RC < \underline{B}_1 \leq RLR^{\infty}$, then $\underline{B}_1C = R{*}\underline{B}_2C$. We can continue this process until we find either

$$\underline{B}C = R{*}^n{*}RC$$

or

$$\underline{B}C = R^{*n} * \underline{D}C \qquad \text{with} \quad \underline{D}C > RLR^{\infty}.$$

In the first case,

$$\mathcal{P}_{\underline{B}C}(\mu) = (1-\mu)(1-\mu^2)\ldots(1-\mu^{2^n})$$

and the only positive root of $\mathcal{P}_{\underline{B}C}(\mu) = 0$ is $\mu = 1$. Thus <u>all itineraries</u> $R^{*n} * RC$ <u>coalesce</u> to RC, since $\underline{K}(g_1) = RC$. The second case above will be discussed later. Next we consider $\underline{D}C > RLR^{\infty}$. Either $\underline{D}C$ is primary, or it can be factorized: $\underline{D}C = \underline{D}_1 * \ldots * \underline{D}_m C$, where $\underline{D}_1 C$ is primary and $\underline{D}_1 C > RLR^{\infty}$ and $\underline{D}_2 C, \ldots, \underline{D}_m C$ are primary (but not equal to C). Then

$$\mathcal{P}_{\underline{D}C}(\mu) = \mathcal{P}_{\underline{D}_1 C}(\mu)\, \mathcal{P}_{\underline{D}_2 C}(\mu^{k_1}) \ldots \mathcal{P}_{\underline{D}_m C}(\mu^{k_1 \cdot k_2 \cdot \ldots \cdot k_{m-1}}) \tag{4}$$

where $k_i = |\underline{D}_i C|$, $i = 1, \ldots, m$. We discuss this case together with the case $\underline{B}C = R^{*n} * \underline{D}C$ described before. There we have

$$\mathcal{P}_{\underline{B}C}(\mu) = (1-\mu)(1-\mu^2)\ldots(1-\mu^{2^{n-1}})\, \mathcal{P}_{\underline{D}C}(\mu^{2^n}) . \tag{5}$$

First of all, no matter whether $\underline{D}C$ is primary or not, the root $\mu_{\underline{D}C}$, $\mu_{\underline{D}C} > 1$, of $\mathcal{P}_{\underline{D}C}(\mu) = 0$ satisfies $\mu_{\underline{D}C} > \sqrt{2}$. Thus if $\underline{B}C = R^{*n} * \underline{D}C$, then by (5), a root $\mu_{\underline{B}C}$ of $\mathcal{P}_{\underline{B}C}(\mu) = 0$ is given by

$$\mu_{\underline{B}C} = (\mu_{\underline{D}C})^{1/2^n}$$

and there is no other root > 1. If $\underline{D}C$ can be factorized, we analyze each factor in (4). Since $\mu > \sqrt{2}$, it follows from what has been said before that none of the factors

$$\mathcal{P}_{\underline{D}_j C}(\mu^{k_1 \cdot \ldots \cdot k_{j-1}})$$

can be zero for $j \geq 2$, because either $\underline{D}_j C > RLR^\infty$ is primary and $\mu^{k_1 \cdot \ldots \cdot k_{j-1}} > \mu^{(Q)}$, $Q = |\underline{D}_j C|$ (since $\mu^{(Q)} < 2$ for all Q and $\mu > \sqrt{2}$). Or $\underline{D}_j C = RC$ and then, $\mathscr{P}_{\underline{DC}}(z) = 1 - z$ but the assumption above shows that $1 - \mu^{k_1 \cdot \ldots \cdot k_{j-1}} \neq 0$.

Thus there remains only the possibility that $\mu_{\underline{DC}} = \mu_{\underline{D}_1 C}$ in (4). All itineraries $\underline{D}_1 * \underline{Q}C$, $\underline{D}_1 C > RLR^\infty$ coalesce in the piecewise linear case and only $\underline{D}_1 C$ (primary) survives. This proves Lemma 4.

So far, we have only discussed finite sequences for g_μ, and we have taken this as an occasion to discuss some aspects of the theory of characteristic polynomials. We shall now adapt the proof of Theorem 1 to the case of the piecewise linear functions g_μ, and see what remains of Lemma 3 for unimodal maps which are not $\mathscr{C}^1$. The net result of this analysis is the

LEMMA III.1.6. Let $\underline{A}$ be a maximal, primary sequence such that $\underline{A} > RLR^\infty$. There is a μ such that $\underline{K}(g_\mu) = \underline{A}$.

Proof. By the definition of maximality, we must have for every finite, maximal $\underline{B}C$ one of the inequalities $\underline{A} > \underline{B} * RL^\infty$ or $\underline{A} < \underline{B} * L^\infty$, since $\underline{A}$ is primary and by Theorem I.2.7. We now repeat the method of the proof of Theorem 1. So again we set

$$M_{\underline{A}} = \{\mu : \underline{K}(g_\mu) < \underline{A}\}$$

and

$$P_{\underline{A}} = \{\mu : \underline{K}(g_\mu) > \underline{A}\} \quad ,$$

and we claim these are open when $\underline{A}$ satisfies the hypotheses of the lemma. Again take $2 > \mu > \sqrt{2}$, with $\underline{D} = \underline{K}(g_\mu) < \underline{A}$. We have to show that for a neighborhood of μ, the inequality is preserved. This poses no problem if at the first index for which $D_i = A_i$, we have $D_i \neq C$. Otherwise, we need the special assumptions of the Lemma. We have $\underline{D} = \underline{B}C$ and either

$\underline{A} > \underline{B}*RL^{\infty}$ or $\underline{A} < \underline{B}*L^{\infty}$. Assume for definiteness the first alternative. An argument similar to Lemma 3 shows that <u>given</u> s, <u>if</u> $\underline{K}(g_{\mu_0}) = \underline{B}C$ <u>then</u> $\underline{K}(g_\mu)$ <u>is one of</u> $\underline{B}C$, $\underline{B}T(\underline{B}U)^s\ldots$ <u>with</u> $T = R$ <u>or</u> L <u>(if</u> $(\partial_\mu g_\mu^n)\big|_{\mu=\mu_0}(0) \neq 0$ <u>with</u> $n = |\underline{B}C|$, <u>both cases occur!) and with</u> $U = R$ <u>if</u> $\underline{B}$ <u>is odd and</u> L <u>if</u> $\underline{B}$ <u>is even, provided</u> μ <u>is sufficiently near to</u> μ_0. (Use the linearity of g_μ away from $x = 0$).

Note that it is the possibility that $T \neq U$ which is different from the case of $\mathscr{C}^1$-unimodal maps, Lemma 3.

EXAMPLE. For $\mu = (1+\sqrt{5})/2$, we have $\underline{K}(g_\mu) = RLC$, while for μ slightly larger, $\underline{K}(g_\mu) = RLL(RLR)^s\ldots$, and for μ slightly smaller, $\underline{K}(g_\mu) = RLR(RLR)^s\ldots$. In particular, as noted in Lemma 4, $RLLRLC = RLC*RC$ lies "in-between" $RLRRLR\ldots < RLLRLC < RLLRLR\ldots$ and does not occur.

We complete the proof of Lemma 6. Since we have assumed $\underline{A} > \underline{B}*RL^{\infty}$ and assuming $\underline{B}$ even, there is an s such that $\underline{A} > \underline{B}R(\underline{B}L)^s\ldots$. Similarly if $\underline{B}$ is odd or $\underline{A}$ satisfies the other inequality.

From this it follows at once that $M_{\underline{A}}$ and $P_{\underline{A}}$ are open, and the remainder of the proof is as in Theorem 1.

<u>Remarks and Bibliography</u>. That the sequence of stable periodic orbits has some regularity, was known to many people experimentally for a large number of one-parameter families of maps. The clearest account was given by Metropolis-Stein-Stein [1973], and explained in the review of May [1976]. The experimental evidence led however many authors to believe that the sequence is always "traversed" in a monotonic fashion. While this seems to be the case for the family $f_\mu(x) = 1 - \mu x^2$, it is manifestly wrong for the family $f_\mu(x) = 1 - 9/4\mu(\mu-4/3)^2 x^2$. The correct statements and proofs have been given in Guckenheimer [1979] and Lanford [1979]. The most thorough study of what happens when f is not $\mathscr{C}^1$ was

done by Derrida-Gervois-Pomeau [1979], especially in connection with the *-product, but part of the results are implicit in Guckenheimer [1979], see also Jonker-Rand [1980].

III.2 ABUNDANCE OF APERIODIC BEHAVIOR

Consider a once differentiable one-parameter family of $\mathscr{C}^1$-unimodal maps as in the preceding section. We assume for definiteness that the parameter varies in [0,1] and for simplicity we also assume the maps f_μ are symmetric functions. If $f_0(1) = 0$ and $f_1(1) = -1$, then all maximal admissible itineraries $\underline{A}$ with $RC \leq \underline{A} \leq RLR^\infty$ occur as the kneading sequence of one of the f_μ's. In particular, many non-periodic kneading sequences can occur, and the discussion of Part II has shown that in some cases one has sensitive dependence on initial conditions, while in other situations one finds an absolutely continuous ergodic invariant measure.

Since the family is $\mathscr{C}^1$-unimodal, we have seen in Lemma III.1.3 that if there is a superstable period for some value of μ (i.e., a finite kneading sequence), then for nearby values of μ there will be still a stable periodic orbit. Thus it follows that the set $\{\mu : f_\mu$ has a stable periodic orbit$\}$ has positive Lebesgue measure. For a qualitative understanding of the "general" nature of one-dimensional maps, viewed as dynamical systems it is of importance to know how frequent the non-periodic behavior is among the members of a one-parameter family.

This can be specified in two directions: a) By measuring the frequency, b) by describing the aperiodic character in more detail.

As an example, it is immediate from the last section, that <u>there is an uncountable set of parameters</u> μ <u>for which</u> f_μ <u>has no stable periodic orbit</u>.

<u>Proof</u>. There is an uncountable set of non-periodic maximal itineraries $\underline{A}$ satisfying $RC \leq \underline{A} \leq RLR^\infty$, namely

$RLR^2 LR^{p_1} LR^{p_2} \ldots$ with $p_i > 2$ for all i. By Lemma 1.3, to each of these there corresponds at least one μ so that the kneading sequence is $\underline{A}$. By Proposition II.6.2, if we assume the f_μ to be S-unimodal then f_μ has no stable periodic orbit. In fact, by choosing the p_i to be sufficiently irregular, we can construct an uncountable set of maximal itineraries which are not periodic and not of the form $\underline{B}*\underline{Q}$. Then Corollary II.7.14 and Theorem II.7.9 imply the corresponding f_μ has sensitive dependence on initial conditions. We have thus proven

PROPOSITION III.2.1. *Let $\mu \to f_\mu$ be a once differentiable map from $[0,1]$ to the space of S-unimodal maps. Assume $\underline{K}(f_0) \le (RLR)^\infty$ and $\underline{K}(f_1) = RLR^\infty$. Then there is an uncountable set of μ for which f_μ has sensitivity to initial conditions.*

The aim of this section is to present a result which goes in the direction of proving the following conjecture: Define $A(\{f_\mu\}) = \{\mu\colon f_\mu$ has no stable periodic orbit$\}$. The conjecture states that under the hypotheses of Proposition 1 one has $\lambda(A(\{f_\mu\})) > 0$ and in addition, $\mu = 1$ is a *point of density* for $A(\{f_\mu\})$, i.e., $\lim_{\mu\to 1} \lambda(A(\{f_\mu\}))/(1-\mu) = 1$. (Jakobson [1978] calls this a Lebesgue point.)

An even stronger conjecture is that the above results hold for the sets $B(\{f_\mu\}) = \{\mu: f_\mu$ has sensitivity to initial conditions$\}$ and $C(\{f_\mu\}) = \{\mu: f_\mu$ has absolutely continuous invariant measure$\}$.

So far results are only known for particular one-parameter families of maps. (B,C in Jakobson [1978], [1979] for $f_\mu(x) = 1 - (1+\mu)x^2$ and similar families and B in Collet-Eckmann [1980] for a family to be described below.)

The remainder of this section is a description--in narratory style--of the proof of Collet-Eckmann [1980]. The reader is referred to the original--very long--paper, if he wants to study the details.

We first comment on our choice of the family f_μ. It will be more useful to choose the parametrization in such a way that the point of density is at $\mu = 0$, not at $\mu = 1$. Furthermore, we shall insist to consider a family of maps with quadratic maximum. Our second motivation comes from a desire to describe a family as close as possible to $1 - (2-\mu)x^2$.

Our family of functions f_μ is defined by

$$f_\mu(x) = \begin{cases} 1 - 2|x| & \text{if } \mu \le |x| \le 1 \\ 1 - \mu - x^2/\mu & \text{if } |x| \le \mu \end{cases} .$$

Although this is not a unimodal family, a simple transformation will bring $f_\mu|_{[\mu-1,1-\mu]}$ to a $\mathscr{C}^1$-unimodal form. Our functions are not S-unimodal, since they are not $\mathscr{C}^3$. But they have all other properties of S-unimodal functions, and in particular their iterates satisfy property R, Section II.4. Thus all conclusions of Sections II.5-7 are, mutatis mutandis, valid for our functions.

We will permanently denote $E_\mu = \{x \mid |x| \le \mu\}$. The graph of f_μ is shown in Figure III.1. Note that f_μ is related to $x \to 4s\,x(1-x)$ with $s = 1 - \pi^2\mu^2/8$ as follows. Define $\hat{f}_\mu(y) = 2/\pi \arcsin(s^{1/2}\sin(\pi y))$. Then $\hat{f}_\mu(y) = 1 - \mu - 2/\mu\, y^2 + \mathscr{O}(\mu^2)$. Thus the relation between f_μ and $\hat{f}_\mu$ is manifest. On the other hand $x \to 4s\,x(1-x)$ is conjugated to $y \to 1 - 2(2s^2-s)y^2$ so that the relation to our standard family is established.

We analyze next the successive iterates $x_n = f_\mu^n(0)$ of the point 0, and in particular the derivative $D_{n,\mu} = d/dx\, f_\mu^n|_{x=f_\mu(0)}$. We shall show that for a large set of μ, $|D_{n,\mu}|$ diverges exponentially, i.e. $|D_{n,\mu}| > 2^{n/7}$. This implies together with the fact that $f_\mu^n(0)$ does not return with a strict period to E_μ, that f_μ has sensitive

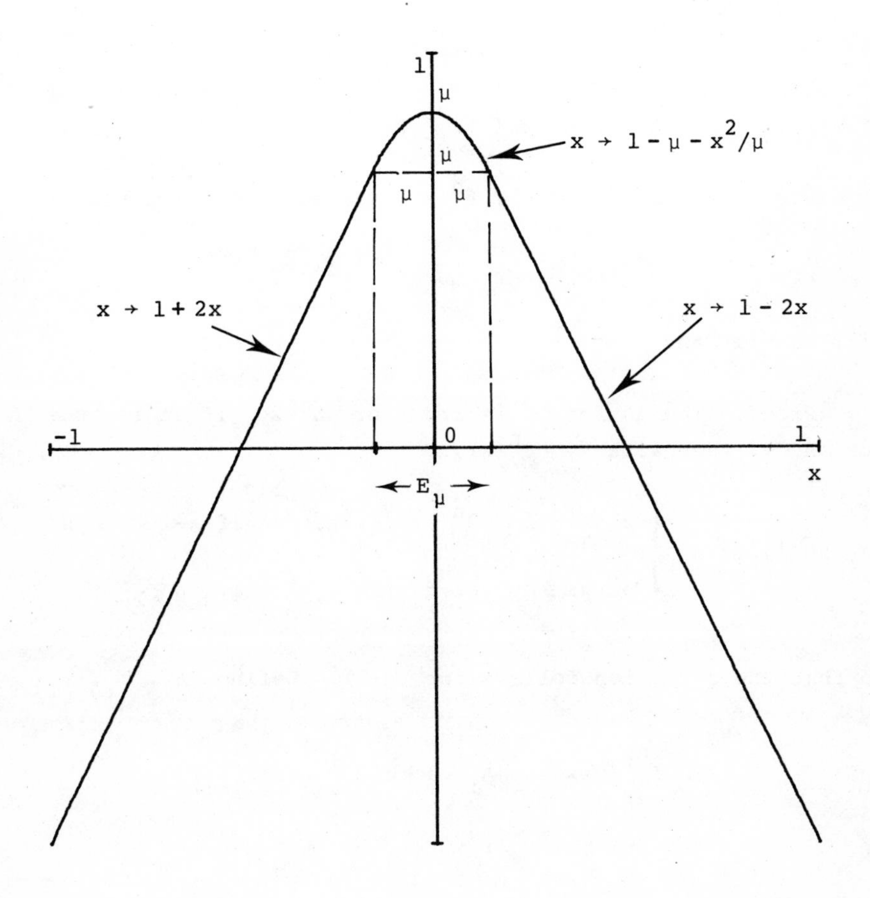

Figure III.1. The function f_μ for $\mu = 0.15$.

dependence with respect to initial conditions in the sense of Section II.7.

Fix a small $\mu > 0$. We analyze the function $f = f_\mu$. Let $x = \tau_{i-1}\mu$, with $|\tau_{i-1}| < 1$. Define n_i to be the smallest value of n for which $|f^n(x)| < \mu$, and set $\mu\tau_i = f^{n_i}(x)$. We shall establish a relation between τ_i, τ_{i-1} and n_i. From the definition of f we have

$$f(x) = 1-\mu(1+\tau_{i-1}^2)$$

$$f^2(x) = 2\mu(1+\tau_{i-1}^2)-1 \quad .$$

We claim that for $2 \le n \le n_i$, $f^n(x)$ can be written in the form

$$f^n(x) = 4A-1+\sigma 2^{n-1}(1+\tau_{i-1}^2)\mu \quad ,$$

with $\sigma = \pm 1$ and $A \in \mathbb{Z}$.

(Proof. The assertion is true for $n=2$. If it is true for $n \ge 2$, then with $y = f^n(x)$,

$$f^{n+1}(x) = \begin{cases} 2(4A)-1+\sigma 2^n(1+\tau_{i-1}^2)\mu, & \text{if } -1 \le y \le -\mu \\ 3-8A-\sigma 2^n(1+\tau_{i-1}^2)\mu \quad , & \text{if } \mu \le y \le 1 \end{cases}$$

so that the assertion follows for $n+1$). Define A_i, σ_i, τ_i by

$$f^{n_i}(x) = 4A_i - 1 + \sigma_i 2^{n_i-1}(1+\tau_{i-1}^2)\mu$$
$$= \tau_i\mu \quad .$$

This also reads

$$\mu\cdot(2^{n_i-1}(1+\tau_{i-1}^2)-\sigma_i\tau_i) = \sigma_i[1-4A_i]$$

or

$$\mu = \frac{B_i}{2^{n_i-1}(1+\nu_{i-1}^2)+\nu_i} \quad , \tag{1}$$

where $B_i = \sigma_i[1-4A_i] \in 2\mathbb{N}-1$ (since $\mu > 0$, and $|\nu_i| < 1$, B_i must be positive) and $\nu_i = -\sigma_i\tau_i$. We are interested in $f^n(0)$, so that we set $\mu_0 = \nu_0 = 0$, i.e.,

$$\mu = \frac{B_1}{2^{n_1-1} + \nu_1} \quad . \tag{2}$$

NOTE: Given $B_i \in 2\mathbb{N}-1$, σ_i and A_i are defined uniquely by

$$\begin{aligned} A_i &= n, \quad \sigma_i = -1 \quad \text{if} \quad B_i = 4n-1 \\ A_i &= -n, \quad \sigma_i = +1 \quad \text{if} \quad B_i = 4n+1 \end{aligned} \quad . \tag{3}$$

The quantities A_i, σ_i, B_i are of course closely related to the kneading sequence of f_μ. The n_i and ν_i are related to the derivative by

$$\left| \frac{d}{dx} f_\mu^m \big|_{x=f_\mu(0)} \right| = \prod_{j=1}^{k} 2^{n_j} |\nu_j| \cdot 2^{m'} \quad , \tag{4}$$

where k is defined by

$$\sum_{j=1}^{n} n_j < m < \sum_{j=1}^{k+1} n_j \quad ,$$

and

$$m' = m - \sum_{j=1}^{k} n_j \quad .$$

(The factor $|\nu_k|$ is absent if $m' = 0$.) Note that all these quantities are functions of μ. Since we are interested in large derivatives, we must be especially careful when a ν_j is near 0. In particular, if $\nu_j = 0$ for $j > 0$, we are in the presence of a superstable periodic orbit for the value of μ in question, and we discard this value of μ, together with a small interval around it. The problem is to choose these intervals sufficiently small so that their union has a relatively small volume, and sufficiently large so that $D_{n,\mu}$ diverges exponentially. The theorems of this section state that these two conditions are compatible. In the study of the excluded volume, we learn a lot about the "typical" behavior of an aperiodic map.

As an example of what is not a "typical" μ, we show that there is an uncountable set of μ for which $f_\mu^n(0) \notin E_\mu$ for all $n > 0$, (so that $|Df_\mu^n(f(0))| = 2^n$), but that this set of μ has Lebesgue measure zero. Namely let $2^{-q-1} \le \mu < 2^{-q}$ be such that the binary representation of $2^q\mu$ does not have $q-2$ consecutive zeros, nor $q-2$ consecutive "1"'s. Then

$$f_\mu^n(0) \quad = \quad 4A_n \pm 2^{n-1}\mu - 1$$

and since $A_n \in \mathbb{Z}$, we have from the condition on μ, that the fractional part of $f_\mu^n(0)$ cannot have more than $q-1$ consecutive zeros, hence $|f_\mu^n(0)| > \mu$. On the other hand, the measure of the set of μ without $q-2$ consecutive zeros or 1's in their binary representation is zero since it is a subset of the numbers without "digit" $\underbrace{0\ldots0}_{q-2}$ and $\underbrace{1\ldots1}_{q-2}$ in their 2^{q-2}-adic representation.

As a first step in describing the excluded volume, we standardize the description of orbits and introduce "resonances" (values of p for which ν_p is very near to zero) and "blocked positions" (the returns to E_μ after p in which the iterated images of 0 and of $x = \tau_p\mu$ have not yet separated).

DEFINITIONS

I. The set $\mathbb{P}$ of primitive resonances depends on μ through the numbers n_i, B_i, ν_i. It is defined by

$$\mathbb{P} = \left\{ p \in \mathbb{N} \,\middle|\, n_{p+1} = n_1 \quad \text{and} \quad B_{p+1} = B_1 \right. \\ \left. \text{and} \quad \nu_p^2 < 2^{-n_1}\Delta^{-4}L^2(p) \right\} \qquad ,$$

where

$$\Delta = |\log \mu| \quad ,$$

$$L(n) = \prod_{j=n+1}^{\infty} (1+j^{-3}) \quad .$$

The intuitive meaning of a resonance is that it is a return to E_μ which is (a) very near to $\nu_p = 0$ and (b) sufficiently near to $\nu_0 = 0$ so that the trajectory to the next return takes the same number of steps $(n_{p+1} = n_1)$ and the same left-right sequence with respect to the maximum $(B_{p+1} = B_1)$. Note that $1 < L(n) < 3.3$.

II. When $p \in \mathbb{P}$, then ν_p is near $\nu_0 = 0$ and this means that we have almost encountered a stable periodic orbit. We shall now devise a test which finds the first $q > p$ for which the orbits starting from ν_p and from ν_0 separate again (provided we exclude a "small" set of μ). By definition, the test $T(p,p)$ is true (passes). Then $T(p,q)$ is recursively defined by

$$T(p,q) \text{ is true if } \Big[n_{q+1} = n_{q-p+1} \text{ and } B_{q+1} = B_{q-p+1} \text{ and } T(p,q-1) \text{ is true and } \operatorname{sign} \nu_q = \operatorname{sign} \nu_{q-p}$$

$$\text{and if } \Big[q \in \mathbb{P} \text{ then } |\nu_q| > |\nu_{q-p}|(1-1/2(q-p)^{-3}) \Big]\Big] \quad .$$

As long as the test passes, we shall say that the two orbits are <u>blocked</u>.

This test says first of all that the two orbits are considered to be blocked if they return simultaneously to E_μ and have the same left-right itinerary. Furthermore if q is a resonance, then $|\nu_q|$ must not be much smaller than $|\nu_{q-p}|$. We shall state, in Equation (6) below that, vaguely speaking, $\nu_q \sim \nu_{q-p}$ when q is a blocked position.

III. We now define the <u>resonance set</u> $\mathbb{P}'$.

1. Let p_0 be the smallest element of $\mathbb{P}$. Then p_0 is the smallest element of $\mathbb{P}'$.

2. If $p \in \mathbb{P}'$, define $t(p)$ as the smallest integer after p for which $T(p,t)$ does not hold, i.e.,

$$t(p) = \inf\{s \mid s > p \text{ and } T(p,s) \text{ is not true}\} .$$

Then the element of $\mathbb{P}'$ following after p is $f(p)$, where

$$f(p) = \inf\{s \mid s \in \mathbb{P} \text{ and } s \geq t(p)\} .$$

As an example:

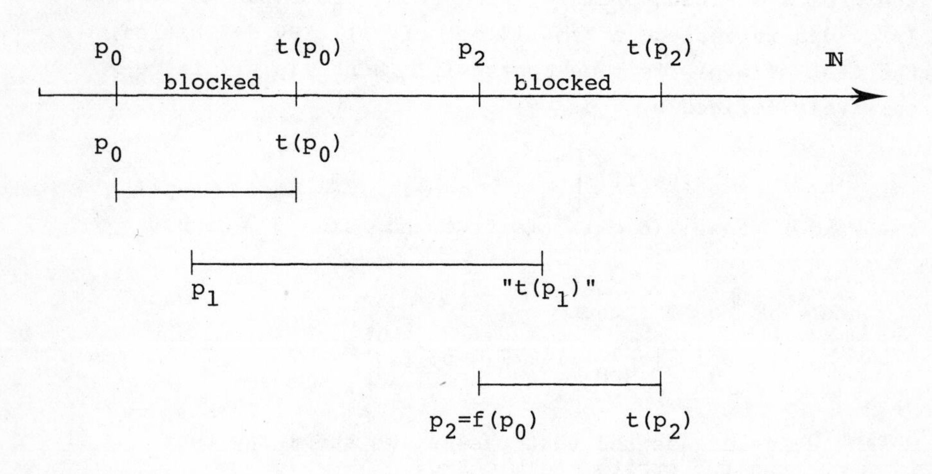

$p_0, p_1, p_2 \in \mathbb{P}$

$p_0, p_2 \in \mathbb{P}', \ p_2 = f(p_0)$.

Figure III.2.

IV. The function q counts the number of "unblocked" positions between $t(p)$ and $f(p)$. The formal definition is

$$q(t) = t \qquad \text{if} \quad t < p_0 \quad ,$$

$$q(t) = q(t-1) \qquad \text{if} \quad p \le t < t(p) \qquad \text{for some} \quad p \in \mathbb{P}' ,$$

$$q(t) = q(t-1) + 1 \qquad \text{if} \quad t(p) \le t < f(p) \qquad \text{for some} \quad p \in \mathbb{P}' .$$

The function r counts the number of $p \in \mathbb{P}$,

$$r(s) = \operatorname{card} \{p \mid p \in \mathbb{P} , \ p < s\} + 1 \quad .$$

V. We next describe five situations in which we exclude a set of μ. Recall that all the quantities B_i, $n_i, \ldots$ are functions of μ and that $\Delta = |\log \mu|$. $\mu_0 > 0$ is chosen in the proofs.

For every $j \in \mathbb{N}$, we define

$$I_j^1 = \Big\{ \mu \,|\, 0 \le \mu \le \mu_0 \quad \text{and} \quad |\nu_j| \le \Delta^{-4j} \quad \text{and}$$
$$(j = 1 \quad \text{or} \quad j \le p_0 \quad \text{or} \quad t(p) \le j \le f(p)$$
$$\text{for some} \quad p \in \mathbb{P}') \Big\} , \qquad j = 1, 2, \ldots \; .$$

Of course, one of the problems will be to describe this exclusion as a function of μ. The idea of excluding I_j^1 is to avoid for the unblocked positions and for $j \in \mathbb{P}'$ and for $j = t(p)$, when $p \in \mathbb{P}'$ that $|\nu_j|$ gets too small. No exclusion is necessary for blocked positions, cf Equation (6). The larger the number of unblocked positions already encountered, the smaller we choose the excluded volume. No exclusion is necessary when j is blocked, because we shall derive from $\nu_j \sim \nu_{j-p}$ (where $p < j < t(p)$) that $|\nu_j|$ does not become too small, since $|\nu_{j-p}|$ does not.

For every $j \in \mathbb{N}$, we define

$$I_j^2 = \Big\{\mu \,|\, 0 \le \mu \le \mu_0 \quad \text{and} \quad 1 - |\nu_j| \le \Delta^{-15j}$$
$$(\text{if } \; j = 1 \; \text{ replace } \; -15 \; \text{ by } -3)\Big\}, \quad j = 1,2,\ldots .$$

This exclusion has the purpose of avoiding ν_j (for all j) to be too near to ± 1. The problem with which this is connected is the following. We would like to argue that when $n_{q+1} \ne n_{q-p+1}$ then $\nu_q - \nu_{q-p}$ is not too small. What may happen, however, is that after n_{q-p+1} steps the image of $\mu\tau_0 = 0$ is near to the boundary of E_μ while $f^{n_{q-p+1}}(\mu\tau_p)$ just very nearly "misses" E_μ. In fact, the exclusion I_j^2 handles the case $n_{q+1} > n_{q-p+1}$, while I_j^3 will deal with the case when $n_{q+1} < n_{q-p+1}$.

$$I_j^3 = \Big\{\mu \,|\, 0 \le \mu \le \mu_0 \quad \text{and} \quad \exists n < n_{j+1}, \; \exists B$$
$$\text{compatible with } \; n, \mu, \; \exists \varepsilon \in \{1,-1\}, \; \text{ such that}$$
$$|\mu(1+\nu_j^2+\varepsilon 2^{-n+1}) - B2^{-n+1}| \le 4\mu 2^{-n}\Delta^{-15j-3}\Big\},$$
$$j = 0,1,2,\ldots .$$

Here, a number B is called <u>compatible with</u> n, μ, if there is an orbit of f_μ for which the pair B,n actually occurs in a first return. The precise definition is

A number $B \in 2\mathbb{N} - 1$ is <u>compatible with</u> n,μ if there is an x, $1 \le x < 2$, such that

1) $|x\mu - B2^{-n+1}| < 2^{-n+1}\mu$,

2) For all $n' < n$ and all $B' \in 2\mathbb{N} - 1$

$$|x\mu - B'2^{-n'+1}| \ge 2^{-n'+1}\mu .$$

We shall now define a further set of μ which will be excluded. Let $\Sigma_{b,s}$ be the sum over the blocked indices $\le s$, i.e., over

$$\{j \mid j \le s \ \text{ and for some } \ p \in \mathbb{P}', \ p \le j \le t(p) - 1\} \ .$$

Then we define

$$I_s^4 = \left\{ \mu \mid 0 \le \mu \le \mu_0 \quad \text{and} \quad \sum_{\substack{j \\ b,s}} n_{j+1} > 10000\ q(s) \log_2 \Delta \right\}, \ s = 1,2,\ldots$$

This set excludes in essence those μ for which a substantial fraction of positions is blocked or for which long returns are blocked. One can prove, that on the complement of I_s^4, one has

$$s < q(s)(1+\varepsilon)$$

for some small $\varepsilon > 0$. Therefore a "typical" orbit is unblocked most of the time. One can also derive the inequality which says that if a resonance occurs after p returns, a "typical" orbit is not blocked for more than p returns. On the other hand, it seems to us that if we single out those μ for which only a finite number of blocked positions occurs, or for which only blocked sequences of bounded length occur, then the Lebesgue measure of these μ is zero.

Our last exclusion is the set I_s^5 (which has in fact zero Lebesgue measure among those μ remaining after the other exclusions). It is essentially

$$I_s^5 = \left\{ \mu \mid 0 \le \mu \le \mu_0 \quad \text{and} \quad s\,\mathbb{N} \subset \{m_i\}_{i=1,2,\ldots}, \quad \text{where} \quad m_i = \sum_{k=1}^{i} n_k \right\} \ , \qquad s = 1,2,\ldots \ .$$

This excludes those μ for which $f^{ks}(0) \in E_\mu$ for $k = 1,2,\ldots$ i.e., for which the image of the maximum periodically returns to E_μ.

This assures that the remaining f_μ have primary kneading sequences (i.e., not of the form $\underline{B}*\underline{Q}$).

RESULTS

Define now $J(\mu_0) = \{\mu \mid 0 \le \mu \le \mu_0$ and $\mu \notin I_s^\ell$, $s = 0,1,2,\ldots$, $\ell = 1,2,3,4,5\}$. Our results are

THEOREM III.2.2. *For $\mu_0 > 0$ sufficiently small, the Lebesgue measure of $J(\mu_0)$ is at least $\mu_0(1-1/\log \mu_0^{-1})$.*

THEOREM III.2.3. *For $\mu_0 > 0$ sufficiently small and for all $\mu \in J(\mu_0)$ one has*

1. $\left|Df^n \Big|_{x=f_\mu(0)}\right| > 2^{n/7}$.

2. f_μ *is topologically conjugate to a piecewise linear map $g_\tau: x \to \tau(1-|x|)$, with $\tau > \sqrt{2}$.*

3. f_μ *has sensitivity to initial conditions.*

We outline the main ideas of the proofs. The first observation is that some minimum time is required between two returns to E_μ; namely

$$2^{n_j} > \mu^{-1}/8 \quad .$$

For the comparison of ν_q with ν_{q-p} when q is blocked, the main observation is:

If $q \notin \mathbb{P}$ then $|\nu_q| \ge 2^{-n_{q+1}/2}\Delta^{-2}$,

(i.e., non-resonant ν's are not arbitrarily small). Also, as long as q is blocked, due to the basic identity (1),

$$\nu_p^2 = \prod_{j=p+1}^{q-1} \frac{1}{2^{n_j-1}(|\nu_j|+|\nu_{j-p}|)} \cdot \frac{|\nu_q-\nu_{q-p}|}{2^{n_{q-1}}} \tag{5}$$

(i.e., ν_p, which is resonant, can be estimated through the square root of $|\nu_q-\nu_{q-p}|$). Using the "blocking," one can show

$$\prod_{j=p+1}^{q-1} |\nu_j/\nu_{j-p}| \sim \Delta^{-4} \quad , \tag{6}$$

and give bounds on $|\nu_q-\nu_{q-p}|$, using the exclusions, namely, e.g., for $p \in \mathbb{P}'$ and $q = t(p)$,

$$|\nu_q - \nu_{q-p}| \geq 2^{-n_{q+1}} \Delta^{-15(q-p+1)} \quad .$$

This serves to bound recursively the derivative as follows. We see that the quantity of interest is (cf (4))

$$R_m = \prod_{j=1}^{m} |\nu_j| 2^{n_{j+1}} \quad .$$

Assume we have already bounded R_{p-1}. Then the main result is that, essentially due to (5),

$$R_{t(p)-1} \sim \Delta^{-8} R_{p-1} \left(2^{n_1} R_{t(p)-p-1}\right)^{1/2} |\nu_{t(p)} - \nu_{t(p)-p}|^{1/2} \quad ,$$

i.e., we have absorbed the potentially dangerous factor ν_p into the last two factors. Using now the bounds on $|\nu_q-\nu_{q-p}|$ one can prove the exponential divergence of R_n.

Next we describe how we bound the excluded volumes. The idea is to measure how fast a resonant interval is traversed when μ varies (very little). The relevant bound shows that

$$\left|\frac{d\nu_s}{d\mu}\right| > \frac{2^{n_1} R_{s-1}}{4\mu}$$

and then we use the formula

$$\int d\mu \sim \int d\nu_s \left|\frac{d\mu}{d\nu_s}\right| \text{ card } [\nu_s^{-1}(\nu_s)]$$

to bound the integrals. One needs a relatively fine decomposition of $0 < \mu < \mu_0$ in order to get the desired bound of Theorem 2.

One uses finally the exclusion of I_s^5 to reduce the discussion to the analysis done in Section II.7 and this yields Theorem 3.2 and 3.3. In particular there is a homeomorphism $\bar{h}$ (which is not necessarily absolutely continuous) such that $\bar{h} \circ f_\delta = g_\tau \circ \bar{h}$.

Remarks and Bibliography. The question of abundance of aperiodic behavior arose naturally in the course of the study of maps on the interval. Experimental results were reported by Lorenz [1979] and Shaw [1978] for maps on the interval, and by Feit [1978] for the Hénon map. A first proof of the fact that aperiodic behavior is not rare for the family $f_\mu(x) = 1 - \mu x^2$ and other explicitly given families was announced by Jakobson [1978], [1979]. He uses the method of induction to show that for many values of the parameter, the induced map is very strongly expanding. He states that this fact, together with the methods of Walters [1975] yields the existence of an absolutely continuous invariant (and ergodic) measure. We were unfortunately unable to reproduce the elegant proof, partly due to the lack of details and partly due to non-mathematical difficulties in communicating with Jakobson. In fact, we have heard most about this work from discussions with Sinai. The work reported here is based on a different method and yields (so far) no absolutely continuous invariant measure. It appeared in Collet-Eckmann [1980].

For an analogous discussion from a topological point of view (in $\mathscr{C}^1$), see Jakobson [1971].

III.3 UNIVERSAL SCALING

We consider families $\mu \to \psi_\mu$ of $\mathscr{C}^1$-unimodal maps which depend in a $\mathscr{C}^1$ fashion on μ, as in Section 1. Then we have seen that if $\psi_\mu(1)$ is near 1 for μ near the left end of the parameter interval and near -1 near the right end, there exists a sequence

$$\mu_1 < \mu_2 < \mu_3 < \dots$$

such that ψ_{μ_j} is superstable of period 2^j and has kneading sequence $R^{*(j-1)}*RC$. It is clear that, if we allow arbitrary (non-monotone) reparameterizations the μ_j's are not unique. In order to establish some unique prescription, we will denote the minimal corresponding values of μ by μ_j and $\lim_{j\to\infty} \mu_j$ by μ_∞.

By investigating numerically a number of one-parameter families, such as $\mu \to 1-\mu x^2$, Feigenbaum [1978, 1979(2)] discovered a striking universality property: For large j, $\mu_\infty - \mu_j$ is asymptotic to

$$\text{const} \times \delta^{-j}$$

where $\delta = 4.66920\dots$ is apparently the same whatever one-parameter family is considered. (Note that, encouragingly, this property of the μ_j's is not changed by making a differentiable change of parameter with derivative which does not vanish at μ_∞.)

Having discovered the universality of δ experimentally, Feigenbaum went on to propose an explanation for it which was inspired by the renormalization group approach to critical phenomena in statistical mechanics. The principal result of this section is to show that Feigenbaum's explanation is correct, at least in a certain limiting regime to be explained below. We will next sketch our version of Feigenbaum's theory, ignoring numerous technical details which will need to be made precise later.

Consider a mapping $\psi \in \mathbb{P}$ where $\mathbb{P}$ denotes the symmetric $\mathscr{C}^1$-unimodal maps and define

$$a = a(\psi) = -\psi(1); \quad b = b(\psi) = \psi(a) .$$

Assume

$$0 < a < b (< 1)$$

and assume also that

$$\psi(b) = \psi^2(a) < a .$$

ψ then maps

$$[-a,a] \text{ onto } [b,1] \text{ and } [b,1] \text{ onto } [-a,\psi(b)] \subset [-a,a],$$

i.e., it exchanges the two non-intersecting intervals $[-a,a]$ and $[b,1]$. Hence $\psi \circ \psi$ maps $[-a,a]$ into itself, and $-\psi \circ \psi$ is again unimodal on $[-a,a]$. (See Figure III.3.) If we reverse orientation and scale up by a factor of $1/a$, i.e. if we make the linear change of variables

$$x_{old} = -ax_{new}$$

then $\psi \circ \psi$ on $[-a,a]$ is transformed into

$$-\frac{1}{a} \psi \circ \psi(-ax) \equiv (\mathscr{T}\psi)(x)$$

on $[-1,1]$.

We will refer to the transformation $\mathscr{T}$ as the <u>doubling transformation</u>. The doubling transformation is essentially just composition of ψ with itself, but combined with restriction of $\psi \circ \psi$ to a subdomain of the original domain and

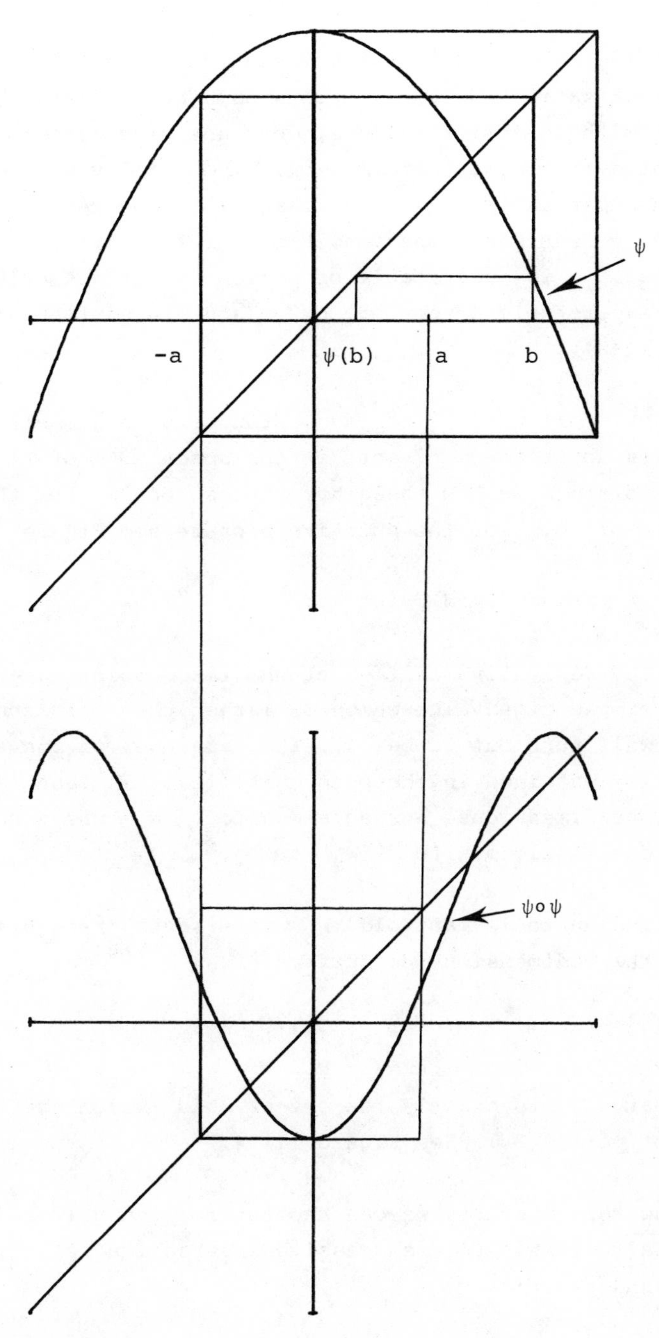

Fig. III.3

then scaling (and reversal of orientation) chosen to preserve the "normalization" $\psi(0) = 1$. This combined operation, in contrast with composition alone, does not give rise to a more complicated-looking transformation. The utility of $\mathcal{T}$ in studying superstable ψ's lies largely in the remark that, provided ψ satisfies the conditions given above for $\mathcal{T}\psi$ to be defined, ψ is superstable of period p if and only if $\mathcal{T}\psi$ is superstable of period $p/2$ (and, in particular, p must be even).

We now, following Feigenbaum, propose some geometrical hypotheses about how $\mathcal{T}$ acts on the space $\mathbb{P}$ of transformations and show how these hypotheses account for the universality of δ. For the putative picture see Figure III.4.

a. $\mathcal{T}$ has a fixed point ϕ.

b. The derivative of $\mathcal{T}$ at the fixed point ϕ has a simple eigenvalue which is larger than one (and which will turn out to be δ); the remainder of its spectrum is contained in the open unit disk. $\mathcal{T}$ thus has a one-dimensional unstable manifold W_u and a codimension one stable manifold W_s at ϕ.

c. The unstable manifold W_u intersects transversally the codimension-one surface Σ_1,

$$\Sigma_1 = \{\psi : \psi(1) = 0\} \quad .$$

(Note that Σ_1 is exactly the set of ψ's which are superstable of period 2.) See Figure III.4.

Using this picture, we can account for the universality of δ as follows: Form successive inverse images $\Sigma_2, \Sigma_3, \ldots$ of Σ_1 under $\mathcal{T}$:

$$\Sigma_j = \mathcal{T}^{-(j-1)}\Sigma_1 \quad .$$

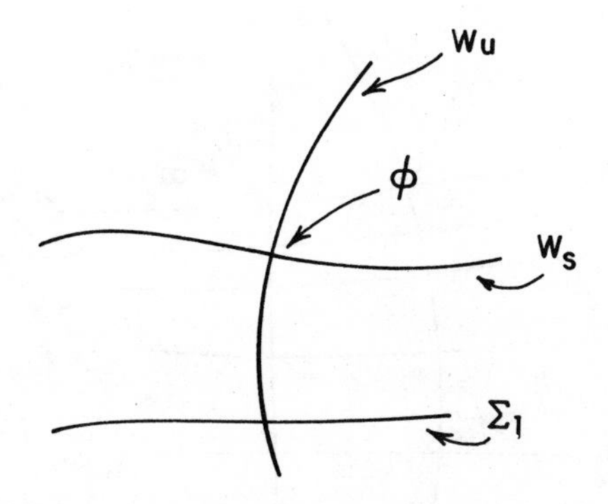

Figure III.4.

Note that if $\psi \in \Sigma_j$ then $\mathscr{T}^{(j-1)}\psi \in \Sigma_1$, so $\mathscr{T}^{j-1}\psi$ is superstable of period 2, so ψ is superstable of period 2^j. The successive Σ_j's come closer and closer to W_s; in fact, a straightforward argument (which we will give in detail later) shows that the separation between Σ_j and W_s decreases exponentially like δ^{-j} for large j, where δ is the largest eigenvalue of the derivative of $\mathscr{T}$ at ϕ.

Now consider a one-parameter family $\mu \to \psi_\mu$ of transformations and regard it as a curve in $\mathbb{P}$. Suppose this curve crosses the stable manifold W_s at $\mu = \mu_\infty$ with non-zero transverse velocity. It is then clear that, at least for large j, there will be a unique μ_j near μ_∞ such that $\psi_{\mu_j} \in \Sigma_j$ (which implies that ψ_{μ_j} is superstable of period 2^j) and that

$$\lim_{j\to\infty} \delta^j(\mu_\infty - \mu_j)$$

exists and is non-zero. See Figure III.5.

Thus, Feigenbaum's hypotheses not only account for the universal rate at which μ_j approaches μ_∞; they also provide in principle an independent prescription for computing δ. They have other consequences as well; we will just mention

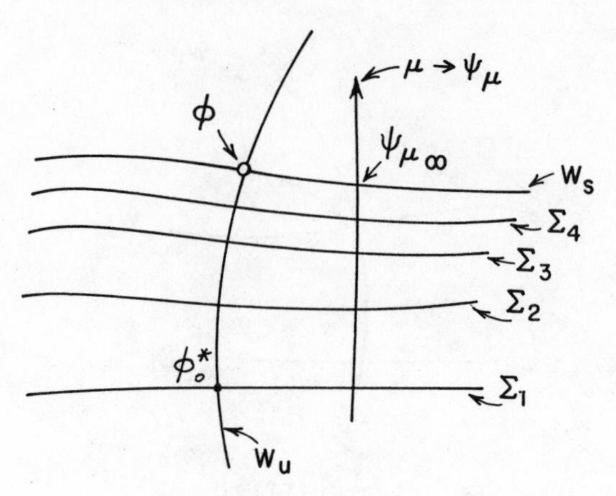

Figure III.5.

two of them here:

1. For all j, $\mathscr{T}^{j-1}\psi_{\mu_j}$ is superstable of period 2. Because $\mathscr{T}$ contracts in the W_s direction, the $\mathscr{T}^{j-1}\psi_{\mu_j}$ converge as $j \to \infty$ to the point of intersection of Σ_1 with W_u, which we will denote by ϕ_0^*. Thus, ϕ_0^* is also universal; for any one-parameter family as above, if we form

$$\psi_{\mu_j}^{2^{j-1}}$$

and scale properly, we get something near to ϕ_0^* for large j. Similarly, $\mathscr{T}^j\psi_{\mu_\infty}$ converges to ϕ.

2. Let $\tilde{\Sigma}_1$ denote the surface

$$\{\psi : \psi^3(1) = -\psi(1)\}$$

(i.e., the set of ψ's such that $-\psi(1)$ is a fixed point for ψ^2). We have seen in Section II.8 that there is an open set of ψ's on $\tilde{\Sigma}_1$ which admit an absolutely continuous invariant measure (and hence which have typical orbits which are not periodic). We will see that $\tilde{\Sigma}_1$ intersects W_u transversally.

(The intersection point will lie <u>above</u> W_s in Figures III.4, 5.) Again form successive inverse images of $\tilde{\Sigma}_1$ under $\mathscr{T}$

$$\tilde{\Sigma}_j = \mathscr{T}^{-(j-1)}\tilde{\Sigma}_1 \quad ;$$

these surfaces converge to W_s, again exponentially with rate δ, from the side opposite to that of the Σ_j's. Again, for each large j, there will be a unique $\tilde{\mu}_j$ near μ_∞ with

$$\psi_{\tilde{\mu}_j} \in \tilde{\Sigma}_j \quad ,$$

and the $\tilde{\mu}_j$'s converge to μ_∞ in the usual way:

$$\lim_{j\to\infty} (\tilde{\mu}_j - \mu_\infty)\cdot\delta^j$$

exists and is non-zero. For large enough j, $\mathscr{T}^{j-1}\psi_{\tilde{\mu}_j}$ will be near to the point of intersection of $\tilde{\Sigma}_1$ with W_u and hence will admit an absolutely continuous invariant measure. From this it is easy to show that $\psi_{\tilde{\mu}_j}$ itself admits an absolutely continuous invariant measure and hence also has orbits which are typically nonperiodic.

Thus μ_∞ is the limit of values of μ for which ψ_μ is chaotic. We warn the reader, however, that not all ψ_μ's for μ just above μ_∞ are chaotic; for example, there is a sequence $\hat{\mu}_j$, again converging to μ_∞ from above, with the same exponential rate, such that $\psi_{\hat{\mu}_j}$ is superstable with period $3\cdot 2^j$.

As indicated earlier, we are going to prove that Feigenbaum's hypotheses are correct in certain cases. As Feigenbaum has noticed, the universality of δ is somewhat relative--its value depends on the function space in which the ψ's are assumed to lie. We will consider functions ψ of the form

$$\psi(x) = f(|x|^{1+\varepsilon})$$

where the function f is analytic in a complex neighborhood of $[0,1]$. We would of course like to deal with the case $\varepsilon = 1$, but the argument we are going to give is a perturbative analysis valid only for sufficiently small positive values of ε. Our results could be expressed in terms of convergent series expansions in ε and various iterated logarithms of ε which are analogues of the ε-expansions occurring in the renormalization-group approach to the theory of critical phenomena. Work in progress, using quite different techniques, indicates that at least partial results can be obtained for $\varepsilon = 1$ (Lanford [1980], Campanino-Epstein-Ruelle [1980]).

We will write $\mathscr{Q}(\Omega)$, or simply $\mathscr{Q}$, for the real Banach space of functions bounded and analytic on Ω, and real on $\Omega \cap \mathbb{R}$, equipped with the supremum norm. For $\varepsilon > 0$, we denote by $\mathbb{P}_\varepsilon (\subset \mathbb{P})$ the set of functions ψ on $[-1,1]$ of the form

$$\psi(x) = f(|x|^{1+\varepsilon})$$

where $f \in \mathscr{Q}$ and satisfying

$$f(0) = 1; \qquad \frac{df}{dt} < 0 \text{ on } [0,1]; \qquad f(1) > -1.$$

We can identify $\mathbb{P}_\varepsilon$ in an obvious way with an open subset of the Banach space $\mathscr{Q}(\Omega)$.

THEOREM III.3.1. <u>For</u> $\varepsilon > 0$ <u>sufficiently small,</u> $\mathscr{T}$ <u>has a fixed point</u> ϕ_ε <u>in</u> $\mathbb{P}_\varepsilon$. <u>If we write</u>

$$\phi_\varepsilon(x) = f_\varepsilon(|x|^{1+\varepsilon})$$

<u>then</u> $f_\varepsilon(t)$ <u>extends to a function jointly analytic in</u> (ε,t) <u>for</u> $\varepsilon \in \{z \in \mathbb{C} \setminus [-\infty,0] : |z| < \varepsilon_0\}$ <u>and</u> $t \in \Omega$. <u>We denote</u> $\phi_\varepsilon(1)$ <u>by</u> λ_ε; <u>then</u>

$$\begin{aligned} \lambda_\varepsilon &= \varepsilon \log \varepsilon + 0(\varepsilon) \\ f_\varepsilon(t) &\doteq 1 - (1-\lambda_\varepsilon)t + 0(\varepsilon^2 \log \varepsilon) \quad . \end{aligned} \tag{3.1}$$

ϕ_ε is an isolated fixed point for $\mathscr{T}$ in $\mathbb{P}_\varepsilon$; it has negative Schwarzian derivative.

Note that, if $\hat{\varepsilon} > \varepsilon$, then $\phi_{\hat{\varepsilon}}(x) = f_{\hat{\varepsilon}}(|x|^{1+\hat{\varepsilon}})$ can also be written as $g(|x|^{1+\varepsilon})$ with g continuously differentiable on $[0,1]$. Thus, $\mathscr{T}$ has at least a one-parameter family of fixed points of the form $g(|x|^{1+\varepsilon})$ with g only once continuously differentiable.

THEOREM III.3.2. The transformation $\mathscr{T}$ is infinitely differentiable in a neighborhood of ϕ_ε in $\mathbb{P}_\varepsilon$. The derivative of $\mathscr{T}$ at ϕ_ε has one simple eigenvalue $\delta_\varepsilon > 2$ which approaches 2 as ε approaches zero. The diameter of the smallest disk centered at zero containing the rest of its spectrum goes to zero with ε.

THEOREM III.3.3. $\mathscr{T}$ has a smooth stable manifold, W_s, of codimension one and a smooth unstable manifold, W_u, of dimension one at ϕ_ε. For each $a \in [-1,1]$ there is a unique point ϕ_a^* on W_u with

$$\phi_a^*(1) = -a \quad .$$

W_u crosses the surfaces Σ_1 and $\tilde{\Sigma}_1$ transversally. Each ϕ_a^* has negative Schwarzian derivative.

THEOREM III.3.4. Let $\mu \to \psi_\mu$ be a continuously differentiable parameterized curve in $\mathbb{P}_\varepsilon$ which crosses the stable manifold W_s with non-zero transverse velocity at $\mu = \mu_\infty$. There exist sequences μ_j and $\tilde{\mu}_j$ converging to μ_∞ from opposite sides such that

$$\lim_{j\to\infty} \delta^j(\mu_\infty - \mu_j) \quad \text{and} \quad \lim_{j\to\infty} \delta^j(\mu_\infty - \tilde{\mu}_j)$$

are both finite and non-zero, and such that ψ_{μ_j} is superstable of period 2^j and $\psi_{\tilde{\mu}_j}$ admits an absolutely continuous invariant measure for each sufficiently large j. Moreover, *the ratio of* $\lim_{j\to\infty} \delta^j(\mu_\infty - \mu_j)$ *to* $\lim_{j\to\infty} \delta^j(\mu_\infty - \tilde{\mu}_j)$ *is also universal, i.e., does not depend on the particular parameterized family under consideration.*

REMARK. One instance of such a parameterized family is

$$\psi_\mu(x) = \psi(\mu \cdot x)$$

for a fixed function ψ sufficiently near to ϕ_ε. We can then in particular take

$$\psi_\mu(x) = 1 - \mu|x|^{1+\varepsilon} .$$

THEOREM III.3.5. *If* $\psi \in W_s$, *then* ψ *has an invariant Cantor set* J.

1. *There is a decreasing chain of closed subsets of* $[-1,1]$

   $$J^{(0)} \supset J^{(1)} \supset J^{(2)} \supset \ldots$$

   each of which contains 0, *and each of which is mapped onto itself by* ψ.

2. *Each* $J^{(i)}$ *is a disjoint union of* 2^i *closed intervals.* $J^{(i+1)}$ *is constructed by deleting an open subinterval from the middle of each of the intervals making up* $J^{(i)}$.

3. ψ *maps each of the intervals making up* $J^{(i)}$ *onto another one; the induced action on the set of intervals is a cyclic permutation of order* 2^i.

We let J *denote* $\cap_i J^{(i)}$. ψ *maps* J *onto itself in a one-one fashion. Every orbit in* J *is dense in* J. *If, besides being on* W_s, ψ *has negative Schwarzian derivative--for which it suffices that it be near* ϕ_ε*--then we have:*

4. *For each* $k = 1,2,\ldots$ ψ *has exactly one periodic orbit of period* 2^{k-1}. *This periodic orbit is repelling and does not belong to* $J^{(k)}$; ψ *has no orbits other than these.*

5. Every orbit of ψ either

   a. lands after a finite number of steps exactly on one of the periodic orbits enumerated in (4).

   or

   b. converges to the Cantor set J in the sense that, for each k, it is eventually contained in $J^{(k)}$.

   There are only countably many orbits of type (a).

THEOREM III.3.6. *Again assume that* $\psi \in W_s$, *and let* $J^{(i)}$, J *be as in Theorem 5. Let* ν *denote the probability measure with support* J *which for each* i *assigns equal weight to each of the* 2^i *intervals making up* $J^{(i)}$.

1. ν *is invariant under the action of* ψ; *it is the only invariant probability measure on* J.

2. *The abstract dynamical system* (ν,ψ) *is ergodic but not weak mixing.*

3. *If* x *is any point of* $[-1,1]$ *whose orbit converges to* J, *and if* f *is any continuous function on* $[-1,1]$, *then*

$$\lim_{N\to\infty} \frac{1}{N} \sum_{n=0}^{N-1} f(\psi^n(x)) = \int_J f \, d\nu \quad .$$

In particular, if ψ is close enough to ϕ_ε so that Theorem 5 holds, then this equality holds for all but countably many x's. Similar results were obtained by Misiurewicz [1978]. The analysis leading to the Cantor set also gives an attractive picture of how the bifurcation at μ_∞ looks.

If $\psi \in \mathscr{D}(\mathscr{T}^n)$, and if $(\mathscr{T}^n\psi)(1) \le 0$, but if ψ is not in W_s then ψ admits a finite decreasing chain

$$J^{(1)} \supset J^{(2)} \supset \ldots \supset J^{(n)}$$

of invariant sets; $J^{(n)}$ is a sort of approximate Cantor set; it is a union of 2^n disjoint closed intervals permuted cyclically by ψ. If in addition ψ is not too far from ϕ, then the space between successive pairs of these intervals contains exactly one periodic point of ψ. These periodic points have periods $1,2,4,\ldots,2^{n-1}$; there is exactly one cycle of each of these periods, and they are all repelling. There are countably many orbits which fall onto one of these repelling orbits after finitely many steps; all others converge to $J^{(n)}$. If we collapse each of the intervals making up $J^{(n)}$ to a point, all such ψ's look the same--they have an attracting periodic orbit of period 2^n together with the simplest set of repelling periodic orbits between them required by simple considerations of connectedness. Each such ψ can thus be thought of as a sort of semi-direct product of the simplest possible ψ which is superstable of period 2^n with the transformation $\mathscr{T}^n\psi$ scaled down and made to act on $J_0^{(n)}$. These $\mathscr{T}^n\psi$'s can of course be very different--e.g., may on the one hand be superstable of period 2 or on the other hand admit an absolutely continuous invariant measure--but the differences act on a small spatial scale and will therefore not be very noticeable for large n. In the limit $n \to \infty$ the approximate Cantor set becomes a true Cantor set which remains attracting and which can crudely be thought of as a single attracting periodic orbit of period 2^∞; at the same time, the spatial scale of the difference between ψ's goes to zero and so the difference disappears entirely.

The remainder of this section consists of proofs (sometimes only sketched) of the results described above.

We first describe the construction of the fixed point, (Propositions 7-10) then sketch the proofs concerning the stable and unstable manifolds (Propositions 11-12) and finally describe the Cantor set.

It is convenient to introduce a new variable α related to ε by

$$\varepsilon = \frac{-\alpha}{1+\log(\alpha)} \quad .$$

Note that for each small positive α there corresponds exactly one small positive ε and vice versa. Any $\psi \in \mathbb{P}_\varepsilon$ can be written uniquely as

$$\psi(x) = f(|x|^{1+\varepsilon}); \qquad f(t) = 1 - t + \alpha t(g(t) - 1) \quad ,$$

with $g \in \mathscr{Q}(\Omega)$.

Working with g rather than ψ is simply a (linear) change of variables in function space. If ψ is given, we will write the g corresponding to $\mathscr{T}\psi$ as $T_\varepsilon g$. The domain of $\mathscr{T}$ is bounded on one side by the surface

$$\psi(1) = 0$$

which corresponds to

$$g(1) = 1 \quad .$$

We are going to show that, for small ε, T_ε is defined and well behaved on the open unit ball in $\mathscr{Q}$ and has a fixed point near zero.

To formulate our results concisely, we need some special terminology. If $\mathscr{X}$ is a normed space and ρ a positive number, we write $\mathscr{X}_\rho$ for the open ball in $\mathscr{X}$ with center 0 and radius ρ. A mapping defined on $\mathscr{X}_1$ will be said to be <u>nearly bounded</u> if it is bounded on each $\mathscr{X}_\rho$ with $\rho < 1$. Similarly, functions will be said to converge <u>nearly uniformly</u> if they converge uniformly on each $\mathscr{X}_\rho$ with $\rho < 1$. These definitions will take care of the factor $1/\psi(1)$ in the definition of $\mathscr{T}$.

PROPOSITION III.3.7. <u>For</u> $\varepsilon > 0$ <u>sufficiently small</u>, T_ε <u>is defined on</u> $\mathscr{Q}_1(\Omega)$. <u>The mapping</u>

$$(\varepsilon, g) \to T_\varepsilon(g)$$

is jointly infinitely differentiable. For fixed ε, derivatives of all orders of T_ε with respect to g are nearly bounded on $\mathscr{Q}_1$. We can decompose T_ε as

$$T_\varepsilon(g) = T_0 g + r_\varepsilon(g)$$

where T_0 is a rank-one linear operator with range the constant functions:

$$(T_0 g)(t) = g(0) + g(1) + g'(1)$$

and r_ε and its g-derivatives of all orders converge nearly uniformly to zero with ε.

COROLLARY III.3.8. 1. For each sufficiently small $\varepsilon > 0$, there is exactly one solution $g_\varepsilon^{(0)}$ for the fixed point problem

$$g = T_\varepsilon(g)$$

in $\mathscr{Q}_{1/2}$. (Here, 1/2 may be replaced by any number less than one.)

$$\varepsilon \to g_\varepsilon^{(0)}$$

is infinitely differentiable and $g_\varepsilon^{(0)}$ approaches zero with ε.

2. $DT_\varepsilon(g_\varepsilon^{(0)})$ varies continuously with ε and approaches T_0 in operator norm as ε approaches zero.

3. Let $0 < \rho < 1$. For sufficiently small ε, the only part of the spectrum of $DT_\varepsilon(g_\varepsilon^{(0)})$ at a distance greater than ρ from 0 is a simple positive eigenvalue δ_ε which approaches two as ε approaches zero. The corresponding eigenspace converges to the space of constant functions.

To prove (1), we write the fixed point problem as

$$g = T_0 g + r_\varepsilon(g)$$

or equivalently as

$$(I - T_0)g = r_\varepsilon(g) .$$

Since $T_0^2 = 2T_0$, we have $(I - T_0)^2 = I$ and so the above equation is equivalent to

$$g = (I - T_0) r_\varepsilon(g) .$$

Since r_ε and Dr_ε converge to zero nearly uniformly with ε,

$$g \to (I - T_0) r_\varepsilon(g)$$

is a contraction on $\mathcal{Q}_{1/2}$ for ε sufficiently small. The existence and uniqueness of $g_\varepsilon^{(0)}$ follows from the contraction mapping principle. The smoothness of the dependence of $g_\varepsilon^{(0)}$ on ε follows from the implicit function theorem in Banach space. (See, for example, Dieudonné [1969].) That $g_\varepsilon^{(0)}$ approaches 0 with ε follows immediately from the nearly-uniform convergence of r_ε to zero.

Part 2 follows from the joint continuity of $DT_\varepsilon(g)$ in g, ε and the continuity of $g_\varepsilon^{(0)}$ in ε. Part 3 follows from 2 by standard perturbation theory (Kato [1966]) and the fact that the spectrum of T_0 reduces to $\{0,2\}$ with 2 a simple eigenvalue whose associated eigenspace is the constant functions.

The proof of Proposition 7 is a relatively straightforward computation supported by some general theorems, for which we refer the reader to the original reference, Collet-Eckmann-Lanford [1980].

The arguments given so far show that the fixed point $g_\varepsilon^{(0)}$ varies smoothly with ε. One can also show that $g_\varepsilon^{(0)}(t)$ is jointly analytic in ε, t. The logarithm appearing in the relation between $\lambda_\varepsilon = -\phi_\varepsilon(1)$ and ε shows that there must be a singularity at $\varepsilon = 0$, and we want to clarify the structure of that singularity.

PROPOSITION III.3.9. $g_\varepsilon^{(0)}(t)$ _is an analytic function of_ $\varepsilon \log \varepsilon$, $1/\log \varepsilon$, $\log(-\log \varepsilon)/\log \varepsilon$, _and_ t. _In particular_, $g_\varepsilon^{(0)}(t)$ _is jointly analytic and bounded in_ ε, t _for_ ε _in a small disk about zero in_ $\mathbb{C}$ _and off the negative real axis and_ t _in_ Ω.

We present a general argument deriving some analytic consequences from the geometric situation we shall encounter. The geometric situation is as follows:

We consider a twice continuously differentiable mapping $\mathcal{T}$ defined on an open set $\mathcal{D}$ in a Banach space $\mathcal{Q}$ and taking values in $\mathcal{Q}$. We do not assume that $\mathcal{T}$ maps $\mathcal{D}$ into itself, but we do assume that it has a fixed point ϕ. We further assume that $D\mathcal{T}(\phi)$, the derivative of $\mathcal{T}$ at ϕ (which is a bounded linear operator on $\mathcal{Q}$) has spectrum which, except for a single simple eigenvalue $\delta > 1$, is entirely contained in the open unit circle. It then follows from invariant manifold theory that $\mathcal{T}$ admits a stable manifold W_s of codimension one and an unstable manifold W_u of dimension one. We will define these submanifolds precisely later; for present purposes, it is good enough to think of W_s as an invariant surface and W_u an invariant curve (the two of them crossing at the fixed point ϕ) with $\mathcal{T}$ acting in a purely contractive way on W_s and in a purely expansive way on W_u.

We also give ourselves two further objects:

> A submanifold Π_1 of $\mathcal{Q}$ of codimension one which intersects W_u transversally at some point $\phi^* \neq \phi$.
>
> A continuously differentiable parameterized curve $\mu \to \psi_\mu$ in $\mathcal{Q}$ which crosses the stable manifold W_s at $\mu = \mu_\infty$ with non-zero velocity.

Although the notation is chosen to suggest the application of the results obtained here to the main topic of this section, the reader should note that these results depend only on a few explicit assumptions about the objects under consideration. Symbols like $\mathscr{T}$, ϕ, W_u, etc. are used in a more general sense here than in the remainder of the section. Moreover, Π_1 will be later on identified with Σ_1 (or with $\tilde{\Sigma}_1$ etc).

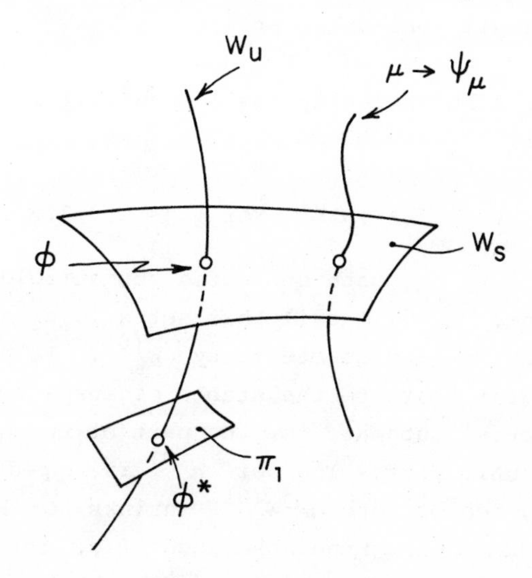

Figure III.6.

From this set-up we want to conclude:

Ca. There exists a sequence μ_n (perhaps defined only for large n), converging to μ_∞, with $\mathscr{T}^{n-1}\psi_{\mu_n} \in \Pi_1$, and such that $\lim_{n\to\infty} \delta^n(\mu_n - \mu_\infty)$ exists and is non-zero.

Cb. The sequence $\mathscr{T}^{n-1}\psi_{\mu_n}$ converges to ϕ^*.

The significance of these conclusions has been discussed above. (We would like to be able to make the more precise assertion that μ_n is the unique value of μ near μ_∞ such that $\mathscr{T}^{n-1}\psi_\mu \in \Pi_1$. Whether this is true or not depends on relatively inaccesible global properties of $\mathscr{T}$; but we can say, informally, that μ_n is the unique such μ for which $\mathscr{T}\psi_\mu, \mathscr{T}^2\psi_\mu, \ldots, \mathscr{T}^{n-2}\psi_\mu$ all lie between W_s and Π_1.)

The first step in our analysis will be to define precisely what we mean by stable and unstable manifolds. This is not entirely routine since, in the application we have in mind, the transformation $\mathscr{T}$ is not invertible; in fact, it is not even locally one-one near ϕ.

If $\mathscr{V}$ is a sufficiently small open ball in $\mathscr{Q}$ with center ϕ then

$$\{\psi \in \mathscr{V} : \mathscr{T}^j\psi \in \mathscr{V} \quad \text{for} \quad j = 1,2,3,\ldots\}$$

may be shown to be a smooth connected submanifold of $\mathscr{V}$ of codimension one. We will call this set a local stable manifold for $\mathscr{V}$ at ϕ and denote it by $W_s^{(0)}$. It passes through ϕ and is tangent there to the stable eigenspace for $D\mathscr{T}(\phi)$, i.e., the spectral subspace for the part of the spectrum which is inside the unit disk. The set $W_s^{(0)}$ is mapped into itself by $\mathscr{T}$, and the sequence of sets $\mathscr{T}^j W_s^{(0)}$ shrinks to $\{\phi\}$, i.e. is eventually contained in any neighborhood of ϕ. The proofs of these facts as well as those to be cited in the next paragraph on local unstable manifolds, can be found in the monograph of Hirsch, Pugh, and Shub [1977].

If we define (again for $\mathscr{V}$ a small open ball with center ϕ)

$$\mathscr{V}_0 = \mathscr{V}; \quad \mathscr{V}_{j+1} = (\mathscr{T}\mathscr{V}_j) \cap \mathscr{V},$$

then $\cap_j \mathscr{V}_j$ is a smooth connected one-dimensional submanifold of $\mathscr{V}$, passing through ϕ and tangent there to the eigenspace of $D\mathscr{T}(\phi)$ corresponding to the large eigenvalue δ. We call this set a local unstable manifold for $\mathscr{T}$ at ϕ and denote it by $W_u^{(0)}$. We have:

$$\mathscr{T}W_u^{(0)} \supset W_u^{(0)} \quad ,$$

and, for any $\psi \in W_u^{(0)}$ and any $j = 1,2,3,\ldots,$ there is a unique $\psi_j \in W_u^{(0)}$ such that

$$\mathcal{T}^j \psi_j = \psi \quad ;$$

moreover, the sequence (ψ_j) converges to ϕ.

The globalization of the stable and unstable manifolds is complicated by the non-invertibility of $\mathcal{T}$. We will simply define what we mean by a stable or unstable manifold without investigating the existence of a unique largest one. Thus we define:

A stable manifold for $\mathcal{T}$ is a smooth codimension-one submanifold W_s of the domain of $\mathcal{T}$ such that:

a. $\mathcal{T} W_s \subset W_s$

b. If $\psi \in W_s$, then $\lim_{j\to\infty} \mathcal{T}^j \psi = \phi$. (Note that this implies that $\mathcal{T}^j \psi \in W_s^{(0)}$ for sufficiently large j.)

c. (Transversality) For any ψ in W_s, the range of $D\,\mathcal{T}(\psi)$ is not contained in the tangent space to W_s at $\mathcal{T}\psi$.

An unstable manifold for $\mathcal{T}$ is a smooth one-dimensional submanifold W_u of $\mathcal{D}$ (not necessarily contained in the domain of $\mathcal{T}$) such that

a. $\mathcal{T}(W_u \cap \mathcal{D}(\mathcal{T})) \supset W_u$

b. If $\psi \in W_u$, there is a sequence ψ_j converging to ϕ such that $\psi = \mathcal{T}^j \psi_j$. (This implies that $W_u \subset \cup_{j=1}^{\infty} \mathcal{T}^j W_u^{(0)}$.)

c. For any $\psi \in W_u \cap \mathcal{D}(\mathcal{T})$, the tangential derivative of $\mathcal{T}$ along W_u at ψ does not vanish.

Since $W_s^{(0)}$ and $W_u^{(0)}$ are, respectively, stable and unstable manifolds, stable and unstable manifolds do exist.

We now need some special terminology. Let Π_j, $j = 1,2,3,...$ and W be submanifolds of $\mathcal{Q}$ of codimension one. We will say that the sequence Π_j converges to W exponentially with rate δ (δ a real number larger than one) if, for each $\psi \in W$ there is a diffeomorphism from $\mathcal{X}_1 \times (-1,1)$, $\mathcal{X}_1$ the open unit ball in some Banach space $\mathcal{X}$, onto a neighborhood $\mathcal{V}$ of ψ (i.e., a set of local coordinates at ψ) such that

1. ψ is the image of $(0,0)$.

2. $W \cap \mathcal{V}$ is the image of $\mathcal{X}_1 \times \{0\}$.

3. For each sufficiently large j, $\Pi_j \cap \mathcal{V}$ is the image of the graph of a mapping $\hat{\Pi}_j : \mathcal{X}_1 \to (-1,1)$

where

4. $\delta^j \hat{\Pi}_j$ converges in the $\mathcal{C}^1$ topology on $\mathcal{X}_1$, to a nowhere vanishing limit.

Intuitively, this means that the separation between Π_j and W varies asymptotically (for large j) like δ^{-j} multiplied by a differentiable function of position on W.

The following proposition is nearly obvious:

PROPOSITION III.3.10. Let Π_j converge exponentially to W with rate δ, and let $\mu \to \psi_\mu$ be a continuously differentiable parameterized curve in $\mathcal{Q}$ crossing W with non-zero transverse velocity at $\mu = \mu_\infty$. There is then a sequence $\mu_j \to \mu_\infty$ (defined for sufficiently large j) such that $\psi_{\mu_j} \in \Pi_j$; the quantity $\delta^j(\mu_\infty - \mu_j)$ converges as $j \to \infty$ to a finite non-zero limit.

Returning to the principal objective of our construction, we see that the proof of Ca is now reduced to constructing appropriately localized pre-images Π_j of Π_1 under $\mathcal{T}^{j-1}$ and showing that they converge exponentially to W_s with rate δ. The following theorem asserts that this is possible; it also asserts that Cb holds.

THEOREM III.3.11. *Let* $\mathcal{T}, \phi, W_s, W_u, \delta, \Pi_1$, *and* ϕ^* *be as above. Then there exists a sequence* (Π_j) *of codimension-one submanifolds of* $\mathcal{Q}$, *converging exponentially to* W_s *with rate* δ, *such that*

$$\mathcal{T}^{j-1}\Pi_j \subset \Pi_1 \quad .$$

Moreover, if $\psi \in W_s$ *and if* $\mathcal{W}$ *is a sufficiently small neighborhood of* ψ *in* $\mathcal{Q}$, *then*

$$\mathcal{T}^{j-1}(\Pi_j \cap \mathcal{W}) \to \{\phi^*\} \quad \text{as} \quad j \to \infty \quad .$$

The first step in proving this theorem is to reduce it to a statement which is local at ϕ. More precisely, it is almost obvious that the theorem as stated is true if we can find an open neighborhood $\mathcal{V}$ of ϕ such that it is true for W_s and W_u replaced by $W_s \cap \mathcal{V}$ and $W_u \cap \mathcal{V}$ respectively, with the added assumption that $\Pi_1 \subset \mathcal{V}$.

Thus, we have only to prove the localized version of the theorem. One does this by choosing special coordinates in which $\mathcal{T}$ and Π_1 take particularly simple forms. We refer to the original paper (Collet-Eckmann-Lanford [1980]) and to Collet-Eckmann [1978] for details. A relatively tedious further analysis makes it possible to prove a global statement about the global unstable manifold, when $\varepsilon > 0$ is small.

PROPOSITION III.3.12. *Let* Ω *be the open disk of radius* 3/4 *and center* 1/2. *There exist constants* B *and* $\varepsilon_0 > 0$,

<u>such that for</u> $0 < \varepsilon < \varepsilon_0$, <u>the unstable manifold for</u> $\mathscr{T}$ <u>in</u> $\mathbb{P}_\varepsilon$ <u>contains a curve</u>

$$a \to \phi_a^*: \phi_a^*(x) = 1 - (1+a)|x|^{1+\varepsilon} - |x|^{1+\varepsilon}(1-|x|^{1+\varepsilon})g_a^*(|x|^{1+\varepsilon})$$

<u>defined on</u> $a \in [0,\tilde{a}]$ <u>and satisfying</u>:

$$\|g_a^*\| \le B\varepsilon \ ; \qquad \left\|\frac{dg_a^*}{da}\right\| \le B\varepsilon; \qquad \phi_{\tilde{a}}^*(\tilde{a}) = \tilde{a} \quad .$$

<u>Moreover if</u> $A(a) = -(\mathscr{T}\phi_a^*)(1)$, <u>then</u> $dA/da > 1.5$ <u>on</u> $[0,\tilde{a}]$.

REMARKS. The estimates developed in this section show that any ψ near ϕ_ε and not on the stable manifold will be driven out of $\mathscr{D}(\mathscr{T})$ by a finite number of iterations of $\mathscr{T}$. This permits to clarify the uniqueness of the μ_j's. Consider the cylinder in $\mathbb{P}_\varepsilon$ corresponding to

$$0 < a < 1; \quad \|g\| \le g_1 \quad .$$

For fixed small ε, and sufficiently small g_1, the stable manifold cuts across this cylinder and thus divides it into two parts. We will refer to the part on the side of $a = 0$ as "above" the unstable manifold and the other part as "below" it. See Figure III.7.

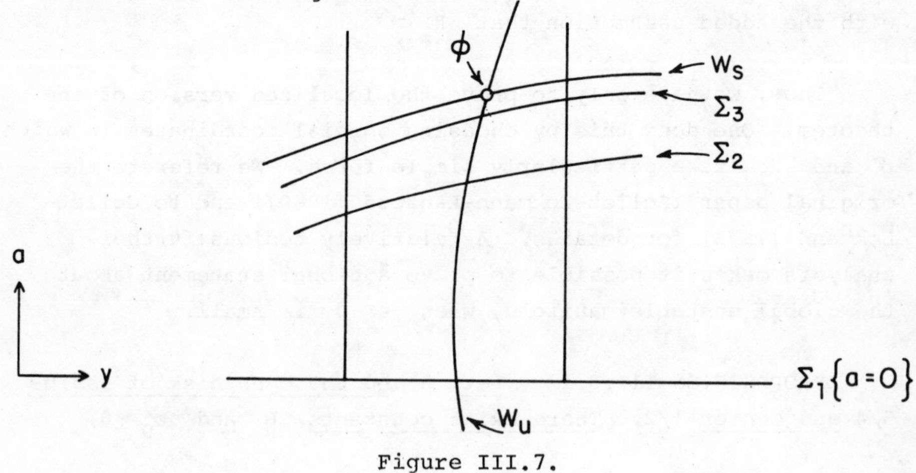

Figure III.7.

The surfaces Σ_j further divide the part of the cylinder above the stable manifold into slabs. If ψ lies between Σ_j and Σ_{j+1}, then $\psi_j \equiv \mathscr{T}^j\psi$ is defined and $(\psi_j)(1) > 0$. Thus, ψ_j maps all of $[-1,1]$ into the interval $[\psi_j(1),1]$, which does not contain 0, and hence $\psi_j^p(0) \neq 0$ for $p=1,2,\ldots$. But ψ_j differs from ψ^{2^j} only by a scale factor, so $\psi^{p2^j}(0) \neq 0$ for all p. This implies that $\psi^p(0) \neq 0$, so ψ is <u>not</u> superstable. Thus: The only superstable ψ's lying in the part of the cylinder above W_s are those on the surfaces Σ_j, $j=1,2,\ldots$. If ψ_μ is a parameterized curve crossing W_s from above when $\mu=\mu_\infty$, at a point inside the cylinder, with non-zero vertical velocity, then for sufficiently large j, the μ_j are uniquely determined by the conditions

$$\psi_{\mu_j} \text{ is superstable of period } 2^j; \quad \mu_j < \mu_\infty; \quad \mu_\infty - \mu_j \text{ is small.}$$

On the other hand, as we have seen in III.1, for large j there are very many values of μ, larger than but near to μ_∞, where ψ_μ is superstable with period 2^j.

In another direction: It is easy to verify (using the implicit function theorem) that, for small ε, g there is a uniquely determined $a=\hat{a}(g)$ such that the ψ corresponding to (a,g) is superstable of period 3. Moreover $g \to \hat{a}(g)$ is smooth and defines a codimension-one surface crossing W_u transversally. Call this surface $\hat{\Sigma}_1$, and apply the theory of the preceding pages to show that its successive inverse images $\hat{\Sigma}_2, \hat{\Sigma}_3, \ldots$ (under $\mathscr{T}$) converge to W_s exponentially with rate δ. If $\mu \to \psi_\mu$ is a parameterized curve as before, there exists a sequence $\hat{\mu}_j$, converging down to μ_∞, such that $\psi_{\hat{\mu}_j} \in \hat{\Sigma}_j$, i.e., $\psi_{\hat{\mu}_j}$ is superstable of period $3\cdot 2^{j-1}$. Moreover, just as before

$$\frac{(\hat{\mu}_j-\mu_\infty)}{\delta} \quad \delta^j$$

converges to a finite non-zero limit. A warning, however: There is another sequence, say $\hat{\hat{\mu}}_j$, with $\psi_{\hat{\hat{\mu}}_j}$ superstable of period $3 \cdot 2^{j-1}$ and $\hat{\mu}_{j-2} > \hat{\hat{\mu}}_j > \hat{\mu}_{j-1}$. In fact, there are infinitely many more distinct, interleaved, sequences of periods $3 \cdot 2^{j-1}$.

Similarly, the equation

$$\psi(a) = a \qquad (\text{i.e.,} \quad \psi^3(0) = \psi^4(0))$$

defines a surface, say $\tilde{\Sigma}_1$, of codimension one crossing W_u transversally. The theory of Section II.8. implies that any $\psi \in \tilde{\Sigma}_1$ sufficiently near to the unstable manifold admits an absolutely continuous invariant measure. It is easy to see, however, that if $\mathscr{T}\psi$ admits an absolutely continuous invariant measure, then so does ψ, so applying the machinery described above, we see that, for each $\mu \to \psi_\mu$ as above, there is a sequence $\tilde{\mu}_j$ converging to μ_∞ such that each $\psi_{\tilde{\mu}_j}$ admits an absolutely continuous invariant measure and such that

$$\delta^j(\tilde{\mu}_j - \mu_\infty)$$

approaches a finite non-zero limit. Many other such examples could be considered, e.g., take for the initial surface the set of ψ's such that $\psi'(x_0) = -1$, where x_0 denotes the unique fixed point of ψ in $[0,1]$. The μ_j's in this case will correspond to bifurcation points, where the orbit of period 2^{j-1} becomes unstable and the orbit of period 2^j appears.

<u>Attracting Cantor Sets</u>

We sketch the construction of J. It is useful to look at Fig. III.3. Let $\psi \in \mathscr{D}(\mathscr{T})$. We write again $a = -\psi(1)$; $b = \psi(a)$; and we will also write $c = \psi(b)$, and we will assume $c \geq 0$. (This would follow automatically if $\psi \in \mathscr{D}(\mathscr{T}^2)$.) Since ψ maps $[-1,1]$ into $[-a,1]$ we may as well restrict its definition to $[-a,1]$. We have seen that ψ maps $[-a,a]$ onto $[b,1]$ and it evidently maps $[b,1]$ in a one-to-one fashion onto $[-a,c]$. Thus, the set $[-a,c]\cup[b,1]$ is mapped onto itself by ψ. Note that this invariant set is constructed out of the original interval $[-a,1]$ by deleting an open subinterval (c,b) in the middle, i.e., as in the first step in constructing a Cantor set. Note also that $[-a,c]$ is mapped onto itself by $\psi\circ\psi$ and that ψ is obtained from the restriction of $\psi\circ\psi$ to $[-a,c]$ by a linear change of variable $x \to -ax$. Observe, finally, also that if $K \subset [-a,c]$ is mapped onto itself by $\psi\circ\psi$, then $J = K \cup (\psi^{-1}K \cap [b,1])$ is mapped onto itself by $\mathscr{T}\psi$.

If $\mathscr{T}\psi$ is also in $\mathscr{D}(\mathscr{T}^2)$, we can apply the same operation to $\mathscr{T}\psi$ and thus obtain an invariant set for ψ by deleting an open subinterval from the middle of each of $[-a,c]$ and $[b,1]$. Continuing, if $\psi \in W_s$ and hence $\psi \in \mathscr{D}(\mathscr{T}^j)$ for all j, we can repeat this operation infinitely often and so obtain an invariant Cantor set for ψ. Let ν be the probability measure on J defined by

$$\nu(J_j^{(i)}) = 2^{-i} \quad \text{for all} \quad i,j$$

(i.e., ν assigns equal weight to each of the intervals making up $J^{(i)}$.) Since

$$\psi^{-1}(J_j^{(i)}) \cap J^{(i)} = J_{j-1}^{(i)} \quad ,$$

the construction of ν implies that it is invariant under the action of ψ_i; on the other hand, this same equation shows that any ψ-invariant probability measure assigning measure zero to the complement of J must assign equal measure to each $J_j^{(i)}$ (i fixed but arbitrary, $j = 0,1,2,\ldots,2^i-1$) and hence must coincide with ν. By standard compactness arguments, there then exists a sequence N_j going to ∞ with j such that

$$\bar{f} = \lim_{j\to\infty} \frac{1}{N_j} \sum_{n=0}^{N_j-1} f\circ\psi^n(x)$$

exists for all continuous functions f on $[-1,1]$ but possibly $\bar{f} \neq \int f d\nu$ for some f. But $f \to \bar{f}$ is a positive linear functional on the space of continuous functions, taking the value 1 on the constant function 1, and vanishing if $f = 0$ on J (since $\psi^n(x) \to J$ by assumption). Thus, there is a probability measure $\bar{\nu}$ on J such that

$$\bar{f} = \int f \, d\bar{\nu} \quad .$$

A standard argument shows that $\overline{f \circ \psi} = \bar{f}$, so $\bar{\nu}$ is ψ-invariant, so $\bar{\nu} = \nu$, contradicting the fact that $\bar{f} \neq \int f \, d\nu$ for some f.

The ergodicity of (ν, ψ) follows at once from the fact that ν is the unique ψ-invariant probability measure on J. On the other hand, the set

$$J \cap J_0^{(1)}$$

is invariant under ψ^2 and has measure 1/2, so ψ^2 is not ergodic, so ψ is not weak mixing. One can also show that the spectrum is discrete.

REMARKS. 1. We note that λ_ε appears as an asymptotic scaling parameter for part of the Cantor set J. Specifically, $J_0^{(n)} \cap J$ and $J_0^{(n+1)} \cap J$ asymptotically differ by a scale factor of $-\lambda_\varepsilon$. This means the following: If we write

$$A_n = a(\psi) a(\mathscr{T}\psi) \ldots a(\mathscr{T}^{n-1}\psi) \quad ,$$

where $\quad a(\psi) = -\psi(1) \quad ,$

then

$$J_0^{(n)} \cap J = A_n J(\mathscr{T}^n \psi) \quad .$$

As $n \to \infty$, $\mathscr{T}^n \psi \to \phi$, so, in a sense which is easy to make precise, $J(\mathscr{T}^n \psi) \to J(\phi)$. Thus,

$$A_n^{-1}(J_0^{(n)} \cap J) \quad \text{and} \quad A_{n+1}^{-1}(J_0^{(n+1)} \cap J)$$

look essentially the same for large n. In other words, $J_0^{(n+1)} \cap J$ looks almost the same as $J_0^{(n)} \cap J$ multiplied by a scale factor of $A_{n+1}/A_n = A(\mathscr{T}^n\psi)$. Again, since $\mathscr{T}^n\psi \to \phi$, this scale factor converges to $a(\phi) = \lambda_\varepsilon$. Observe, however, that this scaling is different for other pieces of the Cantor set. For example, successive terms in the decreasing sequence $J \cap J_1^{(1)} \supset J \cap J_1^{(2)} \supset J \cap J_1^{(3)} \supset \dots$ differ asymptotically by a numerical factor of $\lambda_\varepsilon^{1+\varepsilon}$ rather than λ_ε, and the same is true for any of the sequences $J \cap J_j^{(n)}$ for fixed, non-zero j.

2. It is easy to see that, for ψ near ϕ, $J_0^{(2)}$ is longer than $J_1^{(1)}$. More generally for any n, $J_0^{(n)}$ is the longest of the 2^n intervals making up $J^{(n)}$ and $J_1^{(n)}$ is the shortest. The length of $J_0^{(n)}$ is $A_n(1+a(\mathscr{T}_n))$ which behaves asymptotically like $\text{const}\cdot\lambda_\varepsilon^n$. The conclusion that the longest interval in $J_0^{(n)}$ has length bounded by $\text{const}\cdot \lambda_\varepsilon^n$ holds even for all ψ on W_s since $\mathscr{T}^n\psi$ still converges to ϕ. Feigenbaum [1980] sketches a way to characterize these scalings recursively.

<u>Remarks and Bibliography</u>. Bifurcation diagrams, such as our Figure II.19 are very easy to produce with modern computing equipment. Thus they must have been known to many people. The beauty and regularity of the diagram was very thoroughly discussed in the review article by May [1978]. This review contains also a long list of references. However, it seems that Feigenbaum [1978], [1979(2)] was the first to publish the observation that bifurcations accumulate in a universal, geometric fashion. His papers contain many essential ideas about the reason why this universality holds. The universality seems to have been rediscovered by Coullet-Tresser [1978(1,2)]. Some approximations to calculate the universal constants are described and compared in the paper by Derrida-Gervois-Pomeau [1979]. The first attempt to prove rigorously the existence of the whole scheme sketched by Feigenbaum, was completed in the paper Collet-Eckmann-Lanford [1980]. Some expository

variants have appeared in Collet-Eckmann [1978(1)], [1978(2)], [1980]. These papers deal only with the case of small $\varepsilon > 0$. The difficult case $\varepsilon = 1$ was solved first by Lanford, but is so far unpublished except for the (detailed) announcement Lanford [1980]. Further progress for the case $\varepsilon = 1$ has been made by Campanino-Epstein-Ruelle [1980], who have (at this writing) succeeded in proving existence of a fixed point.

III.4 MULTIDIMENSIONAL MAPS

Infinite sequences of period doubling bifurcations have been observed in higher dimensional systems. One of them is the Hénon map in $\mathbb{R}^2$:

$$\begin{pmatrix} x \\ y \end{pmatrix} \mapsto \begin{pmatrix} 1 - \mu x^2 + y \\ bx \end{pmatrix} .$$

For $b = .3$, the first 11 values of μ for which a doubling bifurcation occurs were computed with good accuracy in Derrida-Gervois-Pomeau [1979].

The aim of this section is to outline a proof of the universal behavior for maps in finite dimensional spaces. We shall assume that the results proven in the preceding section for small ε extend to $\varepsilon = 1$. Lanford [1980] and Campanino-Epstein-Ruelle [1980] reported recently on some decisive progress in this direction and our working hypotheses will be inspired by these results.

Our argument is organized as follows. We first state our hypotheses on the one-dimensional case for $\varepsilon = 1$. We then explain how the renormalization group program can be realized for certain maps on $\mathbb{C}^n$, $n > 1$. This includes the search for a fixed point of a nonlinear transformation and the study of its linearization at this fixed point. A second part of the argument would have to include a detailed description of the stable and unstable manifolds W_s and W_u and of their intersection with submanifolds of codimension one. This argument is not worked out in the literature. It should be similar to that of Section III.3.

<u>Throughout this section maps</u> $\mathbb{C}^n \to \mathbb{C}^n$ <u>are implicitly considered to be real on</u> $\mathbb{R}^n$.

THEOREM III.4.1. There is a map Φ from $\mathbb{C}^n$ to $\mathbb{C}^n$, and a submanifold W_s in the space of analytic functions on $\mathbb{C}^n$ to $\mathbb{C}^n$ (of codimension one and passing through Φ) such that the following is true:

1. Every once continuously differentiable one-parameter family $\mu \mapsto G_\mu$ of analytic maps from $\mathbb{C}^n$ to $\mathbb{C}^n$ which crosses transversally through W_s near Φ has infinitely many bifurcations from a stable period 2^m to a stable period 2^{m+1}, for m sufficiently large.

2. If $\{\mu_m\}$ is the sequence of values of μ for which a period 2^m described in (1) appears, then

$$\lim_{m\to\infty} \frac{\log|\mu_\infty - \mu_m|}{m} = -\log\delta \quad ,$$

where $\delta = 4.669...$ does not depend on the family G_μ.

Before stating our hypotheses, we recall some definitions associated with the one-dimensional problem. Let $\mathcal{M}$ be the set of $\mathscr{C}^1$-unimodal functions for which $g(1) < 0$.

For $g \in \mathcal{M}$ and $\lambda < 0$ we define T by

$$Tg(x) = \lambda^{-1} g \circ g(\lambda x) \quad .$$

Our first set of assumptions is the following:

M1. The equation $Tg = g$ has a solution $\phi \in \mathcal{M}$ which is analytic in some complex neighborhood D_1 of $[-1,1]$, when $\lambda = -.3995...$.

M2. ϕ is symmetric, $\phi(x) = f(x^2)$ and $f'(t) \neq 0$ for $t \in [0,1]$.

M3. For some positive γ, $2\lambda^2 \sup\{|f'(z^2)| : z \in D_1\} < 1 - \gamma < 1$.

M4. f has exactly one zero in $[0,1]$.

The reader can clearly see the analogy of these assumptions with the conclusions of Theorem III.3.1. From M1 we can investigate the derivative of T at the fixed point ϕ. We obtain

$$(DT_\phi h)(z) = \lambda^{-1} h(\phi(\lambda z)) + \lambda^{-1}\phi'(\phi(\lambda z))\, h(\lambda z) .$$

LEMMA III.4.2. <u>If</u> $\sigma(y) = y^n$ <u>for some integer</u> $n \geq 0$

$$\psi_\sigma(x) = -\sigma(\phi(x)) + \phi'(x)\sigma(x)$$

<u>is an eigenvector of</u> DT_ϕ <u>with eigenvalue</u> λ^{n-1}.

<u>Proof</u>. We consider the following family $S(t)$ of maps of the complex plane

$$z \to S(t)z = z + t\sigma(z) ,$$

which is well defined and invertible on a neighborhood of $[-1,1]$ for small t.

If M_λ denotes the operator of multiplication by λ in $\mathbb{C}$, we have

$$T(S(t)^{-1} \circ \phi \circ S(t))$$

$$= M_\lambda^{-1} \circ S(t)^{-1} \circ M_\lambda \circ (M_\lambda^{-1} \circ \phi \circ \phi \circ M_\lambda) \circ M_\lambda^{-1} \circ S(t) \circ M_\lambda$$

$$= (M_\lambda^{-1} \circ S(t)^{-1} \circ M_\lambda) \circ \phi \circ (M_\lambda^{-1} \circ S(t) \circ M_\lambda) .$$

Differentiating with respect to t and setting $t = 0$ we obtain

$$DT_\phi(\psi_\sigma) = \psi_{M_\lambda^{-1} \circ \sigma \circ M_\lambda} ,$$

by using that $\psi_\sigma = \partial_t [S(t)^{-1} \circ \phi \circ S(t)]|_{t=0}$. From this the result follows.

Our last set of hypotheses is the following:

M5. The operator DT_ϕ has a simple eigenvalue $\delta > 1$ which is different from λ^{-1}, λ^{-2}.* The corresponding eigenvector ρ is even. We define $r(x^2) = \rho(x)$.

M6. The eigenvalues δ, λ^{-1}, 1 are the only eigenvalues of modulus ≥ 1. Their corresponding spectral projections are one-dimensional.

Lanford [1980] has essentially completed the proof of M1,...,M4. His method can be extended to prove M5 and M6.

REMARK. In view of Lemma 2 it could seem reasonable to assume that one has found a total set of eigenvectors for DT_ϕ. This is not the case and in fact the family $\{\psi_\sigma : \sigma$ analytic$\}$ has infinite codimension.

We introduce some notations for the n dimensional problem. We will use a fixed decomposition of $\mathbb{C}^n$ into a direct sum $\mathbb{C}^n = \mathbb{C} \oplus \mathbb{C}^{n-1}$. If z is a vector of $\mathbb{C}^n$, its components will be written $(z_0, \underline{z})$. $\|\cdot\|$ will be the norm of $\mathbb{C}^n$ given by

$$\|z\| = \|\underline{z}\|_{\mathbb{C}^{n-1}} + |z_0| \quad ,$$

where $\|\underline{z}\|_{\mathbb{C}^{n-1}} = (\underline{z} \cdot \overline{\underline{z}})^{1/2}$ is the usual norm in $\mathbb{C}^{n-1}$. For an open subset D of $\mathbb{C}^n$ let $\mathscr{H}(D)$ denote the space of analytic and bounded maps from D to $\mathbb{C}^n$. Equipped with the norm

$$\|h\| = \sup \{\|h(z)\| : z \in D\} \ ,$$

*In fact $\delta = 4.6692...$.

this space is a Banach space. We shall mostly consider $\mathscr{H}_\Delta = \mathscr{H}(D(\Delta))$, where $D(\Delta)$ is the convex set

$$D(\Delta) = \{z \in \mathbb{C}^n : \| z - (y_0, \underline{0}) \| < \Delta \quad \text{for some} \quad y_0 \in [-1,1]\}.$$

We now fix a non-zero vector $\underline{\alpha}$ in $\mathbb{C}^{n-1}$ whose norm is bounded by two. Φ will always denote the map

$$z \to \Phi(z) = (f(\zeta(z)), \underline{0}) \quad , \tag{1}$$

where $\zeta(z) = z_0^2 - \underline{\alpha} \cdot \underline{z}$ and where f is the function described in M2. Note that if Δ is sufficiently small, so that $\{\zeta^{1/2}(z) : z \in D(\Delta)\}$ is contained in D_1, then Φ belongs to $\mathscr{H}_\Delta$. The universal behavior asserted in Theorem 1 will be proven in the sequel for one-parameter families of maps which are near to Φ.

It might seem that this is an undue restriction on the one- parameter family. Note however that our problem is invariant under $\mathscr{C}^1$ coordinate transformations. This means that given a one-parameter family $\tilde{G}_\mu$ of maps one might find transformations τ_μ such that $G_\mu = \tau_\mu^{-1} \circ \tilde{G}_\mu \circ \tau_\mu$ satisfies the conditions of Theorem 1. In particular, τ_μ might be constant and change the direction or length of $\underline{\alpha}$. The conclusions of Theorem 1 are then valid for G_μ as well as for $\tilde{G}_\mu$.

We next define a transformation $\mathscr{N}$. Let Λ be the diagonal $n \times n$ matrix given by

$$\Lambda z = (\lambda z_0, \lambda^2 \underline{z})$$

where $\lambda = \phi(1) = -0.3995\ldots$.

LEMMA III.4.3. <u>If</u> Δ <u>is sufficiently small and if</u> G <u>belongs to</u> $\mathscr{H}_\Delta$, <u>and</u> $\|G - \Phi\| < \gamma\Delta$,* <u>then</u> $\Lambda^{-1} \circ G \circ G \circ \Lambda$ <u>also belongs to</u> $\mathscr{H}_\Delta$.

*cf. M3 for the definition of γ.

Proof. From the hypotheses on f and on $u = G - \Phi$ it is easy to verify, using M3, that for $z \in D(\Delta)$,

$$G(\Lambda z) = (f(\lambda^2 \zeta(z)) + u_0(\Lambda z), \underline{u}(\Lambda z))$$

is again in $D(\Delta)$, and therefore $\Lambda^{-1} \circ G \circ G(\Lambda z)$ is well defined. The analyticity follows from that of G and of $G(\Lambda \cdot)$.

DEFINITION. The transformation $\mathcal{N}$ given by

$$\mathcal{N}: G \mapsto \Lambda^{-1} \circ G \circ G \circ \Lambda$$

will be called the renormalization transformation.

Due to Lemma 3, this transformation maps the ball in $\mathcal{H}_\Delta$ centered at Φ and of radius $\gamma\Delta$ into $\mathcal{H}_\Delta$.

LEMMA III.4.4. Φ is a fixed point of $\mathcal{N}$.

Proof. This is an easy consequence of M1, M2, and of the definition of $\mathcal{N}$ and Φ.

According to the general program described in III.3, page 202, we shall now investigate the spectrum of the derivative of $\mathcal{N}$ at Φ. One can easily derive the following expression for the derivative $D\mathcal{N}_G$, where $G \in \mathcal{H}_\Delta$ and $\|G-\Phi\| < \Delta\gamma/2$:

$$(D\mathcal{N}_G h)(z) = \Lambda^{-1}[h(G(\Lambda z)) + DG_{G(\Lambda z)} h(\Lambda z)] .$$

PROPOSITION III.4.5. For sufficiently small Δ, the following assertions hold.

1. $\mathcal{N}$ is a $\mathcal{C}^2$ transformation on $\mathcal{H}_\Delta$ defined in a ball centered at Φ and with radius $\gamma\Delta/2$. $\|D^2 \mathcal{N}_{\Phi+u}(k,h)\| \le \mathcal{O}(1) \|h\| \|k\|$ provided $\|u\| < \gamma\Delta/2$.

2. $D\mathcal{N}_\Phi$ is a compact operator from $\mathcal{H}_\Delta$ into itself.

Proof. We remark that for $\|u\| < \gamma\Delta/2$ and sufficiently small Δ, $(\Phi+u)$ maps the closure of $\Lambda D(\Delta)$ into $D(\Delta)$ by M1 and M3. The assertion (1) is now verified by a direct computation. The compactness of $D\mathcal{N}_\Phi$ is a consequence of Montel's theorem.

We start now an analysis of the spectrum of $D\mathcal{N}_\Phi$. Our result is

THEOREM III.4.6. 1. If $\sigma = (\sigma_0, \underline{\sigma})$ is an analytic map from $\mathbb{C}^n$ to $\mathbb{C}^n$ and if $\Lambda^{-1}\circ\sigma\circ\Lambda = \lambda^m\sigma$ for some integer $m \geq -2$, then the map defined by

$$\Psi_\sigma(z) = -(\sigma_0(f(\zeta(z)),\underline{0}),\ \underline{\sigma}(f(\zeta(z)),\underline{0}))$$

$$+\ (2z_0 f'(\zeta(z))\sigma_0(z) - f'(\zeta(z))\underline{\alpha}\cdot\underline{\sigma}(z),\underline{0})$$

is an eigenvector of $D\mathcal{N}_\Phi$ with eigenvalue λ^m.

2. δ, λ^{-2}, λ^{-1} and 1 are the only eigenvalues of $D\mathcal{N}_\Phi$ of modulus greater or equal to one. The corresponding spectral subspaces are spanned

- for δ by $P = (r\circ\zeta,\underline{0})$, where r is the function defined in M5.
- for λ^{-2}, λ^{-1}, 1 by the vectors Ψ_σ with $\Lambda^{-1}\circ\sigma\circ\Lambda$ equal to $\lambda^{-2}\sigma$, $\lambda^{-1}\sigma$ and σ respectively.

They have the dimensions 1, n-1, n and n^2-n+1.

Proof. 1. This can be shown by a method analogous to that of Lemma 2.

2. It is sufficient to prove the assertions for the restriction of $D\mathcal{N}_\Phi$ to a closed linear subspace which contains $D\mathcal{N}_\Phi\mathcal{H}_\Delta$. The direct sum $\hat{\mathcal{H}}(\Delta) = \mathcal{H}_0(\Delta) \oplus \underline{\mathcal{H}}(\Delta)$ has this property, where

$$\mathscr{H}_0(\Delta) = \{h : h = (h_0, \underline{0}) \in \mathscr{H}_\Delta\},$$

$$\underline{\mathscr{H}}(\Delta) = \{h : h(z) = (0, \underline{h}_1(\zeta(z))) \text{ with } \underline{h}_1 \text{ analytic and bounded on } D_0(\Delta)\},$$

where

$$D_0(\Delta) = \{\zeta(z) : z \in D(\Delta)\}.$$

The restriction of $D\mathscr{N}_\Phi$ to $\hat{\mathscr{H}}(\Delta)$ will be denoted by A. Let $\mathscr{H}' = \mathscr{H}(\Phi(D(\Delta)) \cup D(\Delta))$. The following two lemmas, which we do not prove here, are the main ingredient of the proof that there are no other eigenvectors than those listed in Theorem 6 (cf. Collet-Eckmann-Koch [1980]).

LEMMA III.4.7. _Let_ $(A-\mu)^\nu \Psi_\tau = 0$ _for some_ $\tau \in \mathscr{H}'$ _and some_ $\nu \in \mathbb{N}$. _If_ $\mu \in \{\lambda^k : k = -2,-1,0,\ldots\}$ _then_ $\Psi_\tau = \Psi_\sigma$ _for some_ σ _which is analytic in_ $\mathbb{C}^n$ _and for which_ $\Lambda^{-1} \circ \sigma \circ \Lambda = \mu\sigma$. _Otherwise_ $\Psi_\tau = 0$.

LEMMA III.4.8. _Let_ $u \in \hat{\mathscr{H}}(\Delta)$ _and let_ $(A-\mu)^\nu u = 0$ _for some_ μ _with_ $|\mu| \geq 1$. _Then_ $u = cP + \Psi_\tau$ _for some_ $c \in \mathbb{C}$ _and some_ $\tau \in \mathscr{H}'$.

Using these lemmas, we complete the proof of Theorem 6.2. Let $u \in \hat{\mathscr{H}}(\Delta)$ and in a spectral subspace of $D\mathscr{N}_\Phi$ with eigenvalue μ, $|\mu| \geq 1$. Since $D\mathscr{N}_\Phi$ is compact, and $\mu \neq 0$, this means that for some $\nu \in \mathbb{N}$, $(A-\mu)^\nu u = 0$. By Lemma 8, we conclude that for some c, $u = cP + \Psi_\tau$ for some $\tau \in \mathscr{H}'$. Suppose $\mu = \delta$. Then $(A-\mu)^\nu cP = 0$ and hence $(A-\mu)^\nu \Psi_\tau = 0$, so that by Lemma 7, $\Psi_\tau = 0$. If $\mu \neq \delta$ then $-cP = (\delta-\mu)^{-\nu}(A-\mu)^\nu \Psi_\tau = \Psi_\kappa$ for some $\kappa \in \mathscr{H}'$. From $(A-\delta)^\nu cP = 0$ we have thus $(A-\delta)^\nu \Psi_\tau = 0$ and applying Lemma 7 again we must either have $\Psi_\tau = 0$ or $\mu = \lambda^k$, $k \in \{-2,-1,0\}$ and $u = \Psi_\tau = \Psi_\sigma$ for a σ with $\lambda^{-1} \circ \sigma \circ \Lambda = \lambda^k$. This completes the proof of Theorem 6.

The argument of Section III.3 showed that we should try to find a map T of function space which is of the form

$$TF = \tau_F^{-1} \circ F \circ F \circ \tau_F \quad ,$$

where τ_F is a (nonlinear) coordinate transformation, in such a way that DT, taken at the fixed point has only one unstable eigenvalue, which is simple.

We shall now show that the eigenvalues λ^{-2}, λ^{-1} and 1 of $D\mathcal{N}_\Phi$ can be removed by an appropriate choice of such a new transformation T.

From the definition of Ψ_σ,

$$\Psi_\sigma = \partial_t (I+t\sigma)^{-1} \circ \Phi \circ (I+t\sigma) \Big|_{t=0} \quad ,$$

it can be seen that the eigenvalues λ^n, $n = -2, -1, 0$, correspond to degrees of freedom associated to some change of coordinates, (in particular the eigenvalue 1 corresponds to transformations which are compatible with our choice of the z_0-axis, i.e., which commute with Λ). Since we intend to describe only coordinate independent properties, the eigenvalues λ^n can be eliminated and ultimately play no role in the universal behavior. We shall now work towards the construction of a new transformation whose derivative at the fixed point has spectrum inside the unit circle except for δ.

Let E denote the spectral projection of $D\mathcal{N}_\Phi$ associated to the eigenvalues λ^{-2}, λ^{-1}, 1. The first step is the definition of a map $h \to \sigma[h]$ which satisfies

$$\Psi_{\sigma[h]} = Eh \quad .$$

PROPOSITION III.4.9. *Define* $D' = \Phi(D(\Delta)) \cup D(\Delta)$. *For any* h *in* $\mathcal{H}_\Delta$, *the equation*

$$\Psi_{\sigma[h]} = EH$$

has a unique solution $\sigma[h]$ *in* $\mathcal{H}(D')$. *The map* $h \mapsto \sigma[h]$

<u>is linear and bounded.</u>

<u>Proof</u>. Let $\mathscr{H}$ be the following finite dimensional subspace of $\mathscr{H}(D')$,

$$\mathscr{H} = \{\sigma:\sigma(z) = \nu + z_0\nu' + (0, z_0^2\underline{\mu} + \underline{\mu}'(\underline{z}))$$

$$\text{with } \nu,\ \nu' \in \mathbb{C}^n,\ \underline{\mu} \in \mathbb{C}^{n-1} \text{ and } \underline{\mu}'$$

$$\text{a linear operator from } \mathbb{C}^{n-1} \text{ into itself}\} .$$

It is easy to verify that $\sigma \mapsto Q\sigma = \Psi_\sigma$ is a bounded linear operator from $\mathscr{H}$ to $E\mathscr{H}_\Delta$. By Theorem 6.2 we have $\dim Q\mathscr{H} = \dim E\mathscr{H}_\Delta$. Therefore Q has an inverse Q^{-1} and we can define $\sigma[h] = Q^{-1}Eh$.

We are now able to define our final transformation T. The explicit expression is

$$T:h \mapsto (I+\sigma[D\mathscr{N}_\Phi h])\circ\Lambda^{-1}\circ(\Phi+h)\circ(\Phi+h)\circ\Lambda\circ(I+\sigma[D\mathscr{N}_\Phi h])^{-1} - \Phi.$$

Since for sufficiently small $\|h\|$ the transformation $z \mapsto z + \sigma[D\mathscr{N}_\Phi h](z)$ maps $D(\Delta)$ analytically and one-to-one onto some neighborhood of $D(\Delta/2)$, this transformation T is well defined in some neighborhood of zero in $\mathscr{H}_\Delta$.

The properties of T are summarized in the following theorem.

THEOREM III.4.10. <u>If</u> Δ <u>is sufficiently small, then</u>

1. T <u>is a</u> $\mathscr{C}^2$ <u>transformation from a neighborhood of zero in</u> $\mathscr{H}_\Delta$ <u>to</u> $\mathscr{H}_\Delta$.

2. $DT_0 = (I-E)D\mathscr{N}_\Phi$, <u>where</u> $DT_0 = DT_G$ <u>at</u> $G = 0$.

3. DT_0 <u>is compact and its spectrum consists of the simple eigenvalue</u> δ, <u>and a remainder strictly inside the</u>

unit disk. The eigenvector corresponding to δ is

$$P(z) = (r(\zeta(z)),\underline{0}) \quad .$$

Proof. The theorem is an immediate consequence of our previous results.

The theorem above is the main ingredient for the analysis of universal behavior of maps, which now follows very closely the one of Section III.3. The relevant manifold, apart from W_s, W_u is now the manifold of "maps at their bifurcation point" Σ_0 defined by $\Sigma_0 = \{G-\Phi : G$ has one fixed point in $D(\Delta)$ and DG has one eigenvalue -1 at this fixed point$\}$. In fact, a more careful description of such manifolds is probably necessary in the multidimensional case to make sure that no unwanted maps are retained.

What matters, is that Σ_0 is a piece of a smooth co-dimension-one manifold intersecting transversally the unstable line W_u. Then the analysis can proceed as in Section III.3.

Remarks and Bibliography. The first published observation of the universality for a multidimensional map seems to be Derrida-Gervois- Pomeau [1979]. Soon afterwards, the other results of Figure I.29 were published by Franceschini [1979] and Franceschini and Tebaldi [1979]. None of these authors made any serious attempts to explain the phenomenon in these cases, exceot for the remark that in some sense the Hénon map becomes one-dimensional when iterated sufficiently often. A rigorous description of what this means is then found in Collet-Eckmann-Koch [1980]. Feigenbaum [1979(1)], [1980] published two papers on the power spectrum which one should expect in such situations. His calculations are however only approximate predictions. An interesting conjecture on the case of volume preserving maps has been made by Derrida [1979]. It is easy to check his prediction $\delta = 8.7210...$, $\lambda = -4.0...$, numerically for the case of the Hénon maps. It has also been found by Benettin-Cercignani-Galgani-Giorgilli [1980] for the case of the maps $[0,1]^2 \to [0,1]^2$ defined by

$$\begin{pmatrix} x \\ y \end{pmatrix} \to \begin{pmatrix} x+y \bmod 1 \\ y-\mu f(z+y) \bmod 1 \end{pmatrix} ,$$

where

$$f(x) = x - 4x^3 \qquad \text{for} \quad 0 \le x \le 1/2$$

$$f(x) = (x-1) - 4(x-1)^3 \qquad \text{for} \quad 1/2 \le x < 1 \quad .$$

REFERENCES

R.L. Adler, A.C. Konheim, M.H. McAndrew. 1965: Topological entropy, Trans. Amer. Math. Soc. 114, 309-319 (1965).

G. Benettin, C. Cercignani, L. Galgani, A. Giorgilli. 1980: Universal properties in Conservative Dynamical Systems, Lettere al Nuovo Cimento 28, 1-4 (1980).

L. Block, J. Guckenheimer, M. Misiurewicz, L.S. Young. 1979: Periodic points and topological entropy of one-dimensional maps, Preprint, Proceedings of the International Conference on Dynamical Systems, Northwestern University, 1979, to appear.

R. Bowen. 1975: Equilibrium states and the ergodic theory of Anosov diffeomorphism. Lecture Notes in Mathematics 470, Springer Verlag, Berlin, Heidelberg, New York, 1975.

_______. 1979: Invariant measures for Markov maps of the interval. Comm. Math. Phys. 69, 1-17 (1979).

R. Bowen, J. Franks. 1976: The periodic points of maps of the disk and the interval. Topology 15, 337-342 (1976).

R. Bowen, D. Ruelle. 1975: The ergodic theory of axiom A flows. Inventiones math. 29, 181-202 (1975).

M. Campanino, H. Epstein, D. Ruelle. 1980: Preprint IHES, to appear.

P. Collet, J-P. Eckmann. 1978(1): Bifurcations et groupe de renormalisation. Séminaire d'Analyse, Collège de France (1978).

_______. 1978(2): Renormalization group analysis of some highly bifurcated families. Proceedings of the Mathematical Physics Conference, Bielefeld, 1978.

_______. 1978(3): A renormalization group analysis of the hierarchical model in statistical physics. Lecture Notes in Physics 74, Springer-Verlag, Berlin, Heidelberg New-York, 1978.

_______. 1980(1): On the abundance of aperiodic behaviour for maps on the interval. Commun. Math. Phys. 73, 115-160 (1980).

_______. 1980(2): Properties of continuous maps of the interval to itself. Lecture Notes in Physics 116. Springer-Verlag, Berlin, Heidelberg, New York 1980.

P. Collet, J-P. Eckmann, H. Koch. 1980: Period doubling bifurcations for families of maps on $\mathbb{R}^n$. J. Stat. Phys. (1980), to appear.

P. Collet , J-P. Eckmann, O.E. Lanford III. 1980: Universal properties of maps on an interval. Comm. Math. Physics, (1980), to appear.

P. Coullet, J. Tresser. 1978(1): Itérations d'endomorphismes et groupe de renormalisation. C.R. Acad. Sc., Paris 287, 577 (1978).

______. 1978(2): Itérations d'endomorphismes et groupe de renormalisation. Journal de Physique C5, 25 (1978).

B. Derrida. 1979: Private communication.

B. Derrida, A. Gervois, Y. Pomeau. 1979: Universal metric properties of bifurcations of endomorphisms. J. Phys. A12, 269 (1979).

______. 1978: Iteration of endomorphisms of the real axis and representations of numbers. Annales de l'Institut Henri-Poincaré 29, 305 (1978).

J. Dieudonné. 1969: Foundations of Modern Analysis. Academic Press, New York and London (1969).

M. Dunford, J.T. Schwartz. 1958: Linear Operators. Interscience , New-York (1958).

P. Fatou. 1919: Sur les équations fonctionnelles. Bull. Soc. Math. de France, 47, 161-270 (1919); 48, 33-95, 208-314 (1920).

M. Feigenbaum. 1978: Quantitative universality for a class of nonlinear transformations. J. Stat. Phys. 19, 25-52 (1978), 21, 669-706 (1979).

______. 1979(1): The onset of turbulence. Phys. Letters 74A, 375 (1979).

______. 1979(2): The transition to aperiodic behavior in turbulent systems. Comm. Math. Phys., 1980, to appear.

S. Feit. 1978: Characteristic exponents and strange attractors. Comm. Math. Phys. 61, 249-260 (1978).

V. Franceschini. 1980: Feigenbaum sequence of bifurcations in the Lorenz model. J. Stat. Phys. 22, 397-407 (1980).

V. Franceschini, C. Tebaldi. 1979: Sequences of infinite bifurcations and turbulence in a five-modes truncation of the Navier Stokes equations. J. Stat. Phys. 21, 707-726 (1979).

J. Guckenheimer. 1977: Bifurcations of maps of the interval, Inventiones Math. 39, 165-178 (1977).

______. 1978: Bifurcations of dynamical systems, CIME Lecture (1978).

______. 1979: Sensitive dependence on initial conditions for one-dimensional maps. Comm. Math. Phys. 70, 133-160 (1979).

M.R. Herman. 1979: Sur la conjugaison différentiable des difféomorphismes du cercle à des rotations. Publ. Math. IHES 49, 5-233 (1979).

E. Hille. 1976: Ordinary Differential Equations in the Complex Domain. John Wiley and Sons, New York, London, Sidney, Toronto (1976).

M.W. Hirsch, C.C. Pugh, M. Shub. 1977: Invariant manifolds. Lecture Notes in Mathematics, Vol. 583, Springer-Verlag, Berlin, Heidelberg, New York (1977).

M. Jakobson. 1971: On smooth mappings of the circle into itself. Math. Sbornik 85, 163-188 (1971).

______. 1978: Topological and metric properties of one-dimensional endomorphisms. Dok. Akad. Nauk. SSSR 243, 866 (1978).

______. 1979: Construction of invariant measures absolutely continuous with respect to dx for some maps of the interval. Proceedings of the International Conference on Dynamical Systems, Northeastern University, 1979, to appear, and Comm. Math. Phys. to appear.

L. Jonker, D.A. Rand. 1980(1): A lower bound for the entropy of certain maps on the unit interval. Preprint, Warwick (1980).

______. 1980(2): Bifurcations in one dimension.
I The non-Wandering set
II A versal model for bifurcations (to appear in Inventiones Mathematicae).

______. The periodic orbits and entropy of certain maps of the unit interval (to appear in Proceedings of the London Mathematical Society).

G. Julia. 1918: Mémoire sur l'iteration des fonctions rationelles. J. de Math. Ser. 7, 4, 47-245 (1918).

T. Kato. 1966: Perturbation Theory for Linear Operators. Springer-Verlag, Berlin, Heidelberg, New York (1966).

J.I. Kifer. 1974: On small random perturbations of some smooth dynamical systems. Math. USSR Izvestija 2, 1083-1107 (1974).

Z.S. Kowalski. 1976: Invariant measures for piecewise monotonic transformations. Lecture Notes in Mathematics, Vol. 472, 77-94, Springer-Verlag, Berlin, Heidelberg, New York (1976).

O.E. Lanford III. 1979: Lecture Notes, Zürich, 1979, unpublished.

O.E. Lanford III. 1980: Remarks on the accumulation of period doubling bifurcations. In Mathematical Problems in Theoretical Physics. Lecture Notes in Physics 116, Springer-Verlag, Berlin,Heidelberg, New York (1980).

A. Lasota, J.A. Yorke. 1973: On the existence of invariant measures for piecewise monotonic transformations. Trans. AMS 183, 481-485 (1973).

T. Li, J.A. Yorke. 1975: Period three implies chaos. Amer. Math. Monthly 82, 985-992 (1975).

A. Libchaber, J. Maurer. 1979: Une experience de Rayleigh-Bénard de géométrie réduite; multiplication, accrochage, et démultiplication de fréquences. J. de Physique, Vol. 41, Colloque C3, 51-56 (1980).

E.N. Lorenz. 1979: On the prevalence of aperiodicity in simple systems. Lecture Notes in Mathematics, vol. 755, 53-77, Springer-Verlag, Berlin, Heidelberg, New York (1979).

R.B. May. 1976: Simple mathematical models with very complicated dynamics. Nature 261, 459-467 (1976).

M. Metropolis, M.L. Stein, P.R. Stein. 1973: On finite limit sets for transformations of the unit interval. J. Combinatorial Theory 15, 25-44 (1973).

J. Milnor, P. Thurston. 1977: On iterated maps of the interval, I, II. Preprint, Princeton, 1977.

M. Misiurewicz. 1966: Topological conditonal entropy. Studia Mathematica 55, 175-200 (1976).

______. 1978: Structure of mappings of an interval with zero entropy. Preprint IHES/M/78/249 (1978).

______. 1980: Absolutely continuous measures for certain maps of an interval. Publ. Math. IHES to appear (1980).

M. Misiurewicz, W. Szlenk. 1977: Entropy of piecewise monotone mappings. Astérisque 50, 299-310 (1977).

______. 1980: Entropy of piecewise monotone mappings. Studia Math. 67, 1980, to appear.

M. Morse. 1966: Symbolic dynamics. Mimeographed notes by R. Oldenberger, Institute for Advanced Studies, Princeton, 1966.

S. Newhouse. 1980: The abundance of wild hyperbolic sets and non-smooth stable sets for diffeomorphisms. Publ. Math. IHES 50, 101-152 (1980).

V.I. Oseledec. 1968: A multiplicative ergodic theorem. Ljapunov characteristic numbers for dynamical systems. Trudy Mosk. Mat. Obsc. 19, 179-210 (1968); Translated Trans. Moscow Math. Soc. 19, 197-231 (1968).

W. Parry. 1964: Symbolic dynamics and transformations of the unit interval. Trans. Amer. Math. Soc. 122, 368-378 (1964).

G. Pianigiani. 1979: Absolutely continuous invariant measures for the process $x_{n+1} = Ax_n(1-x_n)$. Bolletino U.M.I. 16, 374-378 (1979).

M.S. Raghunathan. 1980: A proof of Oseledec' multiplicative ergodic theorem. Unpublished.

A. Renyi. 1957: Representations for real numbers and their ergodic properties. Acta Math. Acad. Sci. Hung. 8, 447 (1957).

D. Ruelle. 1977: Applications conservant une mesure absolument continue par rapport à dx sur [0,1]. Comm. Math. Phys. 55, 47-51 (1977).

______. 1978(1): Dynamical systems with turbulent behavior. In Mathematical Problems in Theoretical Physics. Lecture Notes in Physics 80, Springer-Verlag, Berlin, Heidelberg, New York, 1978.

______. 1978(2): An inequality for the entropy of differentiable maps. Bol. Soc. Bras. Math. 9, 83-87 (1978).

______. 1979(1): Sensitive dependence on initial conditions and turbulent behavior of dynamical systems. In bifurcation theory and its applications in scientific disciplines. New York Acad. of Sci. 316 (1979).

______. 1979(2): Thermodynamic formalism. Addison Wesley, Reading, Mass. (1979).

A.N. Šarkovskii. 1964: Coexistence of cycles of a continuous map of a line into itself. Ukr. Mat. Z. 16, 61-71 (1964).

R. Shaw. 1978: Strange attractors, chaotic behavior and information flow. Preprint, University of California at Santa Cruz (1978).

Ya.G. Sinai. 1976: Introduction to Ergodic Theory. Princeton University Press, Princeton, New Jersey, 1976.

D. Singer. 1978: Stable orbits and bifurcations of maps of the interval. SIAM J. Appl. Math. 35, 260 (1978).

P. Stefan. 1977: A theorem of Šarkovskii on the existence of periodic orbits of continuous endomorphism of the real line. Comm. Math. Phys. 54, 237-248 (1977).

S.M. Ulam, J. v. Neumann. 1967: On combinations of stochastic and deterministic processes. Bull. Amer. Math. Soc. 53, 1120 (1947).

P. Walters. 1975(1): Invariant measures and equilibrium states for some mappings which expand distances. Trans. Amer. Math. Soc. 236, 121-153 (1975).

P. Walters. 1975(2): Ergodic theory-introductory lectures. Lecture Notes in Mathematics 458, Springer-Verlag, Berlin, Heidelberg, New York, 1975.

INDEX

LIST OF MATHEMATICAL SYMBOLS

λ	Lebesgue measure
A^- or $\bar{A}$	The closure of the set A
(a,b)	Open interval with endpoints $a < b$
$[a,b]$	Closed interval with endpoints $a < b$
$(a;b)$	Open interval with endpoints a,b (Not necessarily $a < b$)
x'	x a point. x' is defined by $f(x') = f(x)$, $x' \neq x$ whenever this is possible
f^m	$f \circ \ldots \circ f$, m times (f is a function)
f'	Derivative of f (f is a function)
Df^m	Derivative of f^m (f is a function)
$\underline{I}$	A sequence of symbols R, L, C
$\underline{I}(x)$	Itinerary of x
$\underline{I}_f(x)$	Itinerary of x; f is specified to avoid confusion
$\underline{I}_E(x)$	Extended itinerary of x
$\lvert\underline{I}\rvert$	Cardinality of $\underline{I}$
$J(f)$	The interval $[f(1),1]$
$\chi(A)$	The characteristic function of the set A
$\#$	A prime or not a prime
$\mathcal{S}$	Shift operation